SpringerBriefs in Computer Science

SpringerBriefs present concise summaries of cutting-edge research and practical applications across a wide spectrum of fields. Featuring compact volumes of 50 to 125 pages, the series covers a range of content from professional to academic.

Typical topics might include:

- A timely report of state-of-the art analytical techniques
- A bridge between new research results, as published in journal articles, and a contextual literature review
- A snapshot of a hot or emerging topic
- An in-depth case study or clinical example
- A presentation of core concepts that students must understand in order to make independent contributions

Briefs allow authors to present their ideas and readers to absorb them with minimal time investment. Briefs will be published as part of Springer's eBook collection, with millions of users worldwide. In addition, Briefs will be available for individual print and electronic purchase. Briefs are characterized by fast, global electronic dissemination, standard publishing contracts, easy-to-use manuscript preparation and formatting guidelines, and expedited production schedules. We aim for publication 8–12 weeks after acceptance. Both solicited and unsolicited manuscripts are considered for publication in this series.

**Indexing: This series is indexed in Scopus, Ei-Compendex, and zbMATH **

Timothy Kieras • Junaid Farooq • Quanyan Zhu

IoT Supply Chain Security Risk Analysis and Mitigation

Modeling, Computations, and Software Tools

 Springer

Timothy Kieras
New York University
Brooklyn, NY, USA

Junaid Farooq
University of Michigan-Dearborn
Dearborn, MI, USA

Quanyan Zhu
New York University
Brooklyn, NY, USA

ISSN 2191-5768 ISSN 2191-5776 (electronic)
SpringerBriefs in Computer Science
ISBN 978-3-031-08479-9 ISBN 978-3-031-08480-5 (eBook)
https://doi.org/10.1007/978-3-031-08480-5

This Springer imprint is published by the registered company Springer Nature Switzerland AG
The registered company address is: Gewerbestrasse 11, 6330 Cham, Switzerland

To our families whose support has been instrumental in the completion of this work

Preface

Supply chain risk is a well-studied subject in business processes and logistics management literature. However, its scope is evolving and becoming wider as the systems and processes are becoming more complex. Modern information technology (IT), operational technology (OT), and Internet of things (IoT) systems have complex global supply chains. Moreover, there is an intricate blend of software and hardware systems, which are manufactured, controlled, and operated by different entities. It is thus becoming critically important to have knowledge and understanding of what vendors are linked to the system and what risk do these vendors bring to the system operation. The cybersecurity of IoT-enabled infrastructure systems overarchingly depends on the confidentiality, integrity, and availability of the software and hardware components including their supply chain. The complex network of components involves various actors and organizations that design and integrate different sub-components of the larger system. The insecurity of one sub-component in the supply chain can have downstream effects on the security and resiliency of IoT-enabled infrastructure systems.

This book aims to provide the necessary tools for quantitative understanding and assessment of the supply chain risk threats to critical infrastructure owners and operators. In a typical IoT-enabled infrastructure system, there is a complex integration of multiple components enabling various IT and OT functions. Each component is supplied by a vendor or a network of vendors, which have different levels of trustworthiness from the perspective of the stakeholders. Certain suppliers may have a long-standing history of successful operation and comply with essential cybersecurity practices. On the other hand, there are many newer and potentially less secure vendors, which can introduce unknown vulnerabilities to the overall system security. The supply chain front adds another dimension to the system reliability on top of component reliabilities. Furthermore, a particular component in the system may itself be very reliable but may have been procured from a less trustworthy vendor. Similarly, a component may not be very reliable but may have a highly trustworthy supplier. Therefore, it is critically important to understand the delicate interplay between component reliabilities and the trustworthiness of their suppliers.

Currently, there is a severe dearth of supply chain risk assessment tools that prevents system operators to analyze the risk to their infrastructure from a supply chain standpoint. Moreover, there is a lack of tools that can assist with supplier selection from alternatives and provide insights about supply chain decisions. This book is aimed at unfolding the emerging supply chain risk analysis ecosystem and providing a peak into a practical software tool to help analyze the risk. The described software tool, referred to as I-SCRAM, will enable critical infrastructure owners to make risk informed decisions relating to the supply chain while deploying their IT and OT systems. Providing such information to decision-makers will reduce the possibility of being affected by supply chain attacks from malicious IT and OT vendors. We hope that this book will provide a broad understanding of the emerging cyber supply chain security in the context of IoT systems to academics, industry professionals, and government officials.

Brooklyn, NY, USA   Timothy Kieras
Dearborn, MI, USA   Junaid Farooq
Brooklyn, NY, USA   Quanyan Zhu
March 2022

Acknowledgments

We would like to acknowledge the support of our respective institutions, New York University (NYU) and the University of Michigan-Dearborn, that have enabled us to pursue this work. We would appreciate all past and current members of the Laboratory for Agile and Resilient Complex Systems (LARX) at NYU who have provided us invaluable feedback and created an environment to allow intellectually engaging work. Specific thanks go to Yunfei Ge from LARX who has helped finish Chapter 4 with case studies and numerical examples. Without her assistance, this book would not be completed on time. We are also thankful to members of the Center of Cyber Security at NYU, in particular, Prof. Nasir Memon and Dr. Ed Amoroso, who have supported this work from the very beginning. This work is also a result of many unforgettable discussions with our colleagues and friends, to whom we are eternally thankful.

We also acknowledge the instrumental funding support from the Critical Infrastructure Resilience Institute (CIRI), a Department of Homeland Security Center of Excellence at the University of Illinois at Urbana-Champaign. We are thankful for research inputs from Randy Sandone and David Nicol and administrative support from Elaina Buhs and Andrea Whitesell at CIRI. Their continued support has made possible the development of I-SCRAM, a software tool for supply chain risk analysis and mitigation for IT, OT, and IoT systems. We are also grateful to constructive reviews from many anonymous reviewers of this work and insightful comments from many participants who attended our tutorials, workshops, and conference presentations.

Contents

Acronyms

ACC	Adaptive Cruise Control
APT	Advanced Persistent Threat
AROC	Accountability Receiver Operating Characteristic
AUC	Area Under the AROC
BDD	Binary Decision Diagram
BMS	Building Management Systems
CAV	Connected Autonomous Vehicle
CISA	Cybersecurity and Infrastructure Security Agency
CRAC	Computer Room Air Conditioners
CRAH	Computer Room Air Handlers
DCIM	Data Center Infrastructure Management Systems
GPU	Graphical Processing Unit
HTTP	Hypertext Transfer Protocol
HVAC	Heating Ventilation and Air Conditioning
IC	Incentive Compatibility
ICT	Information and communications technology
IoT	Internet of Things
IR	Individual Rationality
IT	Information Technology
LQR	Linear Quadratic Regulator
LRT	Likelihood Ratio Test
MAP	Maximum Aposteriori
NAIC	National Association of Insurance Commissioners
NIST	National Institute of Standards and Technology
OT	Operational Technology
PDU	Power Distribution Units
ROC	Receiver Operating Curve
SCRM	Supply Chain Risk Management
SLA	Service Level Agreement

SMT	Satisfiability Modulo Theories
UPS	Uninterruptible Power Supplies
WHO	World Health Organization

Chapter 1
IoT and Supply Chain Security

Abstract Internet of things (IoT) applications rely on a variety of technological components that are manufactured and operated by different entities around the globe. Supply chain is emerging as the next frontier of threats in the rapidly evolving IoT ecosystem. It is fundamentally more complex compared to traditional information and communications technology (ICT) systems. This chapter highlights potential sources of supply chain risks in IoT systems and their unique aspects along with providing an overview of the fundamental challenges in supply chain risk assessment and mitigation.

1.1 Vendor Landscape of IoT Systems

The Internet of Things (IoT) is being used as a key enabling technology to secure the supply chain of several industries by tracking of assets, raw materials, and supplies. However, the supply chain security of the IoT itself is generally overlooked. The IoT is an interconnection of smart devices and components that come together to provide situational awareness and automated operation of electronic systems [1]. It is not a standalone system obtained from a single supplier or manufacturer, having propriety hardware and software. Instead, it is composed of various interconnected components that may be designed, manufactured, and operated by different entities located in different parts of the world [2].

A generic illustration of the various components along the IoT technology stack and their interconnection is provided in Fig. 1.1. In essence, there are several actors involved in setting up the IoT ecosystem that include sensing/actuating device manufacturers, firmware developers, radio access network service providers, cloud service providers, mobile app developers, and end-users. The endpoint devices are made of embedded hardware that interacts with the physical environment and is driven by software processes referred to as firmware or operating systems. These make use of communication infrastructure, which is composed of access points, gateways, and core IP networks to connect to cloud servers, that in turn host applications and services, which are operated by users via computing devices such as smart phones, smart watches, and voice assistants, etc. More and more systems

Fig. 1.1 A simplified stack of technologies used in IoT applications

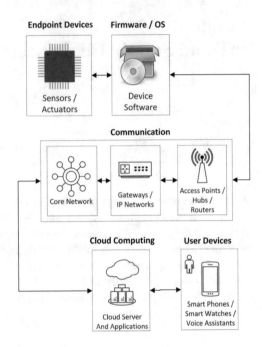

are becoming intelligent and autonomous with the emergence of IoT devices and enhanced ICT infrastructure. However, the integration of multiple devices and components that are designed and manufactured by different entities makes the system extremely vulnerable to cyber-physical attacks [3].

The IoT ecosystem faces serious security threats from traditional attackers due to factors such as low cost, inherent inter-operability, and rapid product development life cycle. Recent large-scale cyber attacks such as *Mirai* [4] botnet and *Stuxnet* [5] have exploited some of the vulnerabilities in IoT systems. Apart from other traditional cyber and physical threats to the IoT ecosystem, the supply chain is emerging as a new source of potential threats. It emanates from the fact that the IoT systems are often deployed in a decentralized manner where managers acquire and deploy equipment needed for improving the efficiency of their business lines. It is generally done without investigating who the suppliers actually are and what risks the suppliers bring to the overall system. Small enterprises, in general, do not have the necessary resources to manage or even assess their risks by increasing automation via the IoT [6]. Therefore, besides tackling traditional cyber-physical threats to the IoT, steps need to be taken to ward off threats at the supply chain front as well. This chapter provides an analysis of some of these threats, research challenges, and potential future directions. We restrict our focus on the supply chain dimension of attacks and risks in the IoT arena.

1.2 Brief Taxonomy of Supply Chain Security

Supply chain security has traditionally been linked to logistics management and ensuring continuity of supplies in industrial processes. Its scope, however, is growing broader as it becomes more critical to understand vendor involvement in operating and maintaining systems and infrastructures that depend on interconnected electronic devices. Recent attacks and data breaches involving ICT suppliers such as Huawei and SolarWinds have raised important concerns about vendors for overall system security [7, 8]. Typically, the supply chain involves individual nodes and links as shown in Fig. 1.2 and hence, supply chain security can broadly be categorized into the following two major concerns. One is the security of supply chain links or third-party interconnections,and the other one is the security of supply nodes or third-party productions.

Security of the Supply-Chain Links The links of the supplier networks as depicted in Fig. 1.2 represent the cyber, physical, or logical dependencies between two suppliers. One intermediate supplier depends on the provision of services or products from another supplier. The link captures, for example, the physical process of transportation between two suppliers, the cyber process of informational provision or IT support, and the logical process in which one product is integrated into another product through designs and production. The security of the supply chain links involves maintaining the continuity of supplies from a supplier to a procurer in a timely manner to ensure meeting end-user requirements. Towards this end, supply chain security implies preventing disruption of the supply chain links because of unforeseen circumstances, natural disasters, or terrorist attacks, etc. One example is the security of medical/pharmaceutical supplies [9] due to COVID-19 because of labor shortages, adulteration, contamination, etc. In the context of link-level supply chain security, the associated Confidentiality, Integrity, and Availability (CIA) triad are described as follows:

- **Confidentiality:** Breaches in security may result in privacy leakage of transactions between suppliers and buyers. The goal of supply chain security here is to keep the confidentiality of the users and the source of the products. To protect the

Top-level Supplier   Malicious Supplier   Intermediate Supplier   End-User

Fig. 1.2 Typical supply chain networks involving nodes and links: A node represents the product suppliers. A link represents the cyber, physical, or logical dependencies of one supplier (or end-buyer) on the other suppliers. The product of an intermediate supplier relies on the transportation and the integration of the products from top-level suppliers. Potential supply-chain attacks may occur both at individual nodes (illustrated by the red color malicious node), supplier links (illustrated by the red color link between the intermediate supplier and end-user), or a combination of the two

confidentiality of the link transactions including information of the supplier and the buyers, blockchain technologies have been shown to be promising solutions [10, 11].

- **Integrity:** The products provisioned by the suppliers in the supply chain may be contaminated by attackers during the process of transmission or transportation. A potential solution to ensure verification of supplied products is based on blockchain technologies [11, 12].
- **Availability:** Any link in the supply chain network can be disrupted due to reasons such as shortage of labor, or delays in component deliveries. To avoid the adverse effects of such events, effective planning and diversification of vendors is needed [13, 14].

Security of the Components Produced or Designed by the Third Party or Beyond Each node of the supply chain network depicted in Fig. 1.2 presents a supplier and its associated products. The security issues can occur at the node level; i.e., it is essential to ensure that products obtained from suppliers are free from flaws, defects, and malicious functionalities. An example is the presence of backdoor channels in microchips or IT equipment produced by third party vendors. Recent supply chain attacks such as SolarWinds [15] and Target data breaches [16] are prime case studies. In the context of node-level supply-chain security , the CIA-triad are described as follows:

- **Confidentiality:** Third-party malicious products collect user's data and private information as we have seen in the infamous Target data breach in 2013 [16].
- **Integrity:** The design of the third party product is intentionally flawed and produces a product that does not satisfy the specifications or has hidden unknown vulnerabilities that can cause malfunctioning as we have seen in SolarWinds attacks [15].
- **Availability:** The design of third-party product leads to a functional breakdown or disruption of operations in the system of the users. For example, the supply-chain induced ransomware attack can lead to the closure of a manufacturing plant as seen by Honda's recent hit by ransomware [17].

Systemic Supply Chain Security The security of the entire supply chain network involves ensuring the system level security from a combination of unreliable supplier links and risky components in the supply chain. In a complex integration of components and system process flow, the security risks propagate through the system and system level security solutions are needed. For instance, a particular vendor in the supply chain may be unreliable or malicious but may not have a critical role in system security. On the other hand, a particular supply chain link may be compromised and thus may result in system malfunction despite un-compromised individual supplier nodes. Hence, the security of components in the supply chain and the security of supply chain links is intertwined. Node-level security leads to link-level security and link-level security affects node-level security.

A holistic approach is needed to ensure supply chain security since any particular type of solutions are insufficient on its own. For instance, the verification of the source can reduce the man in the middle attacks and hence make the next party safer. However, the supply chain links also need to be secure to pass on the benefits down the supply chain for achieving end-to-end supply chain security. Therefore, individual components in the supply chain need to be hardened along with improving the link-level security. This book focuses on making assessments of systemic cyber risk using information about individual node vulnerabilities and the system interconnects to understand how risk flows in IoT systems and networks. The analysis is then to suggest remedial actions in terms of cost-effective vendor selection and diversification of the supply chain.

1.3   IoT Supply Chain Risk: Hard to Observe and Hard to Control

Supply chain risk has long been a matter of great concern for businesses and corporations. In fact, supply chain risk management (SCRM) is a standard functional area across many industries such as consumer goods, food, industrial products, etc., and is considered to be a vital component in securing revenues and profitability of enterprises. Security of information and communications technology (ICT) equipment has been an area of immense focus in recent times. since more advanced and sophisticated methods have emerged to attack IT/OT systems. Cyber-physical attacks on these systems may result in significant monetary and non-monetary losses. To counter threats from such attacks, the U.S. National Institute of Standards and Technology (NIST) has prepared a comprehensive list of best practices for SCRM in traditional ICT systems [18].

The development and growth of the IoT are further enhancing security concerns. Although the flexibility of communication and interaction between devices results in tremendous benefits, however, it also opens doors for attackers and malicious actors to sabotage the system. With the emergence of vendor-based attacks and the involvement of global players, there are rising concerns about the security of the IoT supply chain. The IoT is a special class of ICT systems and is evolving rapidly. The interconnection of systems and devices enables a much richer attack surface as opposed to traditional ICT systems. Moreover, the supply chain of the IoT is extremely complex, globally distributed, and highly inter-connected. In addition, the IoT is still a grossly unregulated technology in terms of security standards unlike food, where the risks are better understood. It is mainly because the ecosystem is highly diverse and the consequences of attacks are relatively unknown. In certain industries such as food and medicine, there are agencies that regulate the safety standards. It is because the risk assessment has been done by testing the products repeatedly on subjects and evaluating the results. However, in the IoT ecosystem, there are limitless functionalities as well as possibilities of malfunction and malicious activity. Hence, determining the possible attacks and enumerating the consequences becomes extremely challenging.

In summary, the risk landscape of the IoT supply chain is extremely diverse. The suppliers may use backdoor channels of devices, inject viruses, provide faulty chips, or load with malicious software. These are merely a tip of a vast number of possibilities that the IoT systems can be attacked. The alarming concern is that these IoT systems are set to control national critical infrastructure resources as well as improve battlefield effectiveness. The supply chain risks are hard to observe and harder to control. The risk propagates from one device to the other and gets amplified as the IoT ecosystem becomes more complex. It is not straightforward to determine where to regulate the entire system.

1.3.1 Dissecting Supply Chain Links in IoT

There is a delicate interplay between suppliers and devices in an IoT ecosystem. To illustrate the different types of interactions that may be present between suppliers and devices, we provide an example in Fig. 1.3 where there are two devices obtained from two different suppliers. While the supply chain can be constructed several levels deep due to individual components in devices being manufactured by different entities, however, for the sake of simplicity, the immediate supplier of standalone devices is considered. In such a scenario, the following different interactions between the supply chain actors might be present.

- **Device-Supplier Interactions:** This is a typical buyer-supplier interaction. The devices are procured from the suppliers and have service contracts including maintenance, upgrades, security patches, etc. The devices have security and support requirements that need to be met under the agreements.
- **Supplier-Supplier Interactions:** Suppliers may have different front-end companies but common connections at the back end. This is typically common in the tech world where corporations have mergers and takeovers. Different suppliers may be owned by a common entity having more control over the supply chain of the IoT network. A nexus of supply chain actors may result in the possibility of coordinated attacks using backdoor channels and other forms of advanced persistent threats.
- **Device-Device Interactions:** These interactions are present due to the inter-connectivity of the IoT devices to provide desired functionality [19]. These interactions are significant since they allow supply chain risks from one device to transfer to the other independently of its own supply chain.

A more detailed illustration of the supply chain interaction with the physical IoT network is provided in Fig. 1.4. There is a component graph that defines the underlying connectivity of devices that make up the IoT ecosystem. Each component has its independent supply chain. However, the supply chains of devices may be linked not only via external affiliations but also via the physical connectivity

Fig. 1.3 Key interactions between different players in the supply chain ecosystem of the IoT

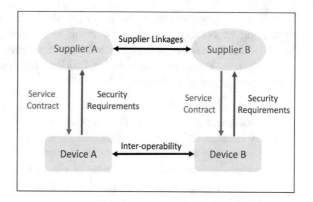

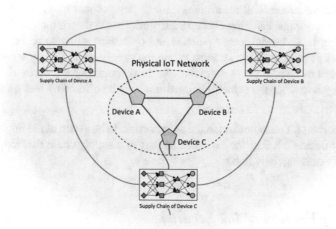

Fig. 1.4 Mapping of IoT and the underlying supply chain networks

of devices in the IoT network. It implies that the risk reciprocates among devices in the network. In other words, my risk becomes your risk and your risk becomes my risk. This makes the analysis of supply chain risks in IoT systems extremely convoluted.

1.4 IoT Risk Implications and Consequences

The impact of risk in IoT is critically important to analyze since it deals with the physical world and any attack or malicious activity may result in significant consequences such as physical damages, operational disruption, and danger to human safety (e.g., see case studies in [20–25]). For instance, a malfunctioning of heating or cooling systems may result is sudden power surges resulting in breakdowns. It may cause a significant loss of revenue as well as damage to the power system. Therefore, there are implications and consequences of the risk that can be categorized as follows:

- *Monetary Implications:* The risk inevitably translates to monetary impact since any disruption or damage to infrastructure would lead to loss of revenue and/or safety hazards. Therefore, the financial impact is important to take into account while selecting the supply chain of IoT networks.
- *Legal Implications:* In the occurrence of a large-scale cyber incident, the liability and responsibility need to be determined. It is important to determine to what extent are the supply chain actors liable for security breaches and what actions can be taken against them. Therefore, there is a need to map out the liability network.
- *Policy Implications:* Risk determines a lot of policies that must be followed by the IoT ecosystem. Patching and upgrade policies are determined by how risky the system is. Cyber-insurance policies and premiums will also depend on how much risk is present in the system. While the supply chain is only a part of the total risk, however, it plays a crucial role in determining the cyber risk of the overall system since the same functionalities may be offered by less trustworthy suppliers.

It is important for stakeholders in the supply chain to be aware of the implications so they can decide whether they want to be part of a supply chain that may render them legally or financially liable.

1.4.1 Key Features of IoT Security

The IoT itself and its security are drastically different than traditional ICT systems. Firstly, there are many different players participating in an unregulated system. Then, the devices are highly inter-operable allowing for limitless possibilities for applications. In fact, it is up to the individual users to build their own desired functionalities and enforce policies on the system. Unlike the Internet, there is no standard protocol stack for the IoT ecosystem. This makes it difficult to embed security into the protocols. The inherent inter-operability in the IoT creates security challenges and vulnerabilities. To this end, the role of the supply chain in IoT is also completely different. While existing wisdom in SCRM for ICT does act as a useful guideline, it may not be sufficient to tackle the more complex nature of IoT networks and the associated supply chain [26]. A summary of the key differences in IoT systems and their security as compared to conventional ICT systems is provided below:

- *Interaction with the physical world:* The IoT devices interact with the physical world with the aid of actuation capabilities as opposed to conventional mobile and computing systems. It results in completely different consequences compared to ICT systems since it may endanger human safety, damage equipment, or cause operational disruptions.
- *Limited access control and management:* The IoT devices are often low-powered with limited computational capabilities. The complete access and management functionalities may not be built into these devices.
- *Different cyber-security standards:* The security and privacy requirements for IoT device operation may be completely different from conventional ICT systems due to different authentication and access mechanisms.
- *Heterogeneous ownership and de-centralized control:* There is no network administrator that has control over the IoT device configurations. Unlike routers and IP networks, the network administrator may not even have a complete view of all the connected devices in the network.

1.5   Challenges in Cyber Supply Chain Risk Analysis of IoT

Since the IoT is inherently a decentralized system, it is difficult to exert control over the entire supply chain. However, the challenges go much beyond the regulation of the supply chain [27]. It is important to study and analyze the threat ecosystem in the IoT landscape. It implies that the potential sources of attack are identified and their potential implications studied in terms of the functionality and/or damage caused to the overall system. In this aspect, some challenges are related to technical aspects of IoT, while others emanate from the logistics and analysis or decision-making standpoint [28]. Some of the key logistical challenges are as follows:

- *Lack of control over upstream supply chain:* There is no control over the upstream supply chain from a device owner's point of view. In other words, the buyers do not have complete information about the cyber-physical supply chain of the products.
- *Disclosure of supply chain information:* Not all suppliers are ready to clearly articulate their cyber security practices and disclose their supply chain information. Some of it is obvious due to privacy reasons and competitor-sensitive information.
- *Awareness of vulnerabilities:* The suppliers of IoT equipment may not be fully aware of all the possible vulnerabilities in their products. This makes it harder to determine the possible attack paths and analyze risk.
- *Centralized database of vulnerabilities:* There is no centralized database of known vulnerabilities and attacks that can serve as a guideline to identify risks and possible attacks.

- *Heterogeneous supply chain management practices:* The management practices for supply chain to mitigate associated security risks are diverse and depend on the industry. heterogeneity across application sectors.

Apart from the logistical constraints, there are also technical challenges in managing the security of the IoT devices that are presented as follows:

- *Lack of management controls:* centralized network management may not be available for the IoT. There is a need for developing management platforms to provide more control to the administrator over the IoT infrastructure.
- *Inflexible hardware:* IoT device hardware may not be serviceable, meaning that it cannot be repaired, customized, or inspected internally.
- *Heterogeneous ownership:* The devices are owned and operated by separate entities resulting in less control over policy implementation.

Finally, some of the decision-making and policy questions that need to be addressed are as follows:

- *Risk informed procurement and deployment:* The decisions to procure and deploy IoT network devices need to be done in a risk-informed manner to allow for cost-benefit analysis.
- *Contingency planning:* The IoT network requires an arrangement of contingencies since suppliers may end security updates or discontinue support for the equipment.
- *Risk-Conscious supplier contracts:* Contracts for installation services should include risk as an essential factor to enable a more secure infrastructure.

Tackling these challenges by finding out novel ways to counter them is important for the research community and policymakers. These are some of the opportunities for researchers and technologists to come up with ways to counter these different types of challenges that will pave the way for securing the IoT ecosystem from supply chain threats.

1.6 Supply Chain Resilience

In a complex system of systems setting with many supply chain actors and their convoluted interactions, it is an extremely challenging problem to mitigate and control supply chain-oriented risks. The approach to tackle the problem is to first completely understand the ecosystem from a supply chain viewpoint and then take appropriate measures to control the risks. In light of the highlighted challenges, there are two potential approaches that can be followed to tackle the problem. The first is the top-down approach, which is centralized, while the second one is the bottom-up approach, which takes a de-centralized viewpoint.

1.6.1   Top-Down Approach to Managing Risk

Governments and policymakers are unable to micro-manage and control individual users of technology to adopt certain practices, particularly in the technology supply chain front. A top-down approach uses a regulatory view of controlling supply risks in the IoT ecosystem. At the outset, policies and restrictions can be imposed on certain supply chain actors. For instance, certain suppliers of equipment may be banned for use in industry due to detected malicious practices or excessive testing and standards may be enforced on certain suppliers based on their trust and reliability levels. Furthermore, they can be mandated to have compulsory disclosure of vulnerabilities to form a centralized database of threats. These will eventually lead to the imposition of tariffs and security requirements on the suppliers. Once the policies are in place, the hope is that the managers and users of IoT technology will be aware of the risks they import by procurement from certain suppliers. This will ultimately result in risk-aware decisions by the users of technology based on other considerations such as cost and functionality. In essence, using a top-down approach, policy drives the underlying technology and supply chain actions. The hope is that centralized awareness and decision-making may have a trickle-down effect to secure the IoT ecosystem from the supply chain threats. Eventually, it might lead to the development of secure supply chain architectures for IoT [29] ecosystems. An illustration of the main stages in the top-down approach is provided in Fig. 1.5.

Fig. 1.5 Top-down approach for managing supply chain risks

1.6.2   Bottom-Up Approach to Managing Risk

The bottom-up approach uses a totally different view from the top-down approach. It aims to map out the view first and then lead to the development of policies to control the risks. The first step is to study and analyze the threat landscape, i.e., sources of attack and potential impacts in terms of functionality and anticipated loss/damage caused. This enables the formation of a comprehensive view of threats and vulnerabilities that are both adversarial and non-adversarial. Once the view has been mapped out, there needs to be a more holistic and integrated measurement of risk. Compound metrics for analyzing the risk as well as the impact are needed. The risk is generally measured as the impact times the likelihood. While the likelihood can be determined using attack trees, the impact needs to be studied more carefully by examining the inter-dependencies and information flows. Then, the goal is to develop mitigation strategies. New infrastructure can then be developed such as a centralized management platform that is in control of the administrators to have a network-wide view of the IoT ecosystem and the supply chain actors involved. While platforms may not be required for individual home users of IoT devices, however, enterprises may need to have a comprehensive tool that allows them to have a clear map of their deployments and the associated risks and propagation. Once this is done, then policies and best practices can be developed for wider dissemination and enforcement. An example of such policy guidelines is the strategic principles that have been proposed by the U.S. Department of Homeland Security for securing the IoT [30]. Consequently, road maps for implementation can be developed by individual industries according to their requirements [31]. The bottom-up approach can be summarized in the flow shown in Fig. 1.6. In essence, the technology and risk assessment drive the development of policy and regulations.

Fig. 1.6 Bottom-up approach for managing supply chain risks

1.7 Overview of the Book

Due to the comparatively unconstrained nature of supply chain threats, potentially posed by a malicious or compromised supplier, risk analysis must shift from a vulnerability-centered approach to the modeling of suppliers and components as a system. Chapter 2 introduces the supply chain risk model for IoT systems and analysis techniques. While attack tree techniques provide a foundation for our methodology, these techniques are adapted to include suppliers and supplier groupings. The inclusion of supplier trust is grounded in the role played by suppliers in risk analysis procedures. Borrowing from established methods in reliability analysis, the use of minimal cutsets and importance measures provide measures of risk across a system and its individual components. The results of the risk assessment are then used to inform the decision-making process and aid the selection of suppliers when alternatives are available. Chapter 3 provides cost effective strategies to diversify the supply chain for reducing systemic risk. Finally, Chap. 4 provides a systematic approach toward accountability investigation of supply chain attacks resulting in appropriate cyber insurance contracts to mitigate threats from vendors in the system. Finally, Chap. 5 provides an overview of a computational tool that is customized for cybersecurity risk analysis and mitigation in IoT-enabled infrastructure systems.

References

1. J. Farooq, Q. Zhu, *Internet of Things-Enabled Systems and Infrastructure* (Wiley, 2021), ch. 1, pp. 1–8. [Online]. Available: https://onlinelibrary.wiley.com/doi/abs/10.1002/9781119716112.ch1
2. J. Farooq, Q. Zhu, *Resource Management in IoT-Enabled Interdependent Infrastructure* (Wiley, 2021), ch. 2, pp. 9–13. [Online]. Available: https://onlinelibrary.wiley.com/doi/abs/10.1002/9781119716112.ch2
3. J. Farooq, Q. Zhu, *Network Defense Mechanisms Against Malware Infiltration* (Wiley, 2021), ch. 8, pp. 97–124. [Online]. Available: https://onlinelibrary.wiley.com/doi/abs/10.1002/9781119716112.ch8
4. C. Kolias, G. Kambourakis, A. Stavrou, J. Voas, DDoS in the IoT: Mirai and other botnets. Computer **50**(7), 80–84 (2017)
5. R. Langner, Stuxnet: Dissecting a cyberwarfare weapon. IEEE Secur. Privacy **9**(3), 49–51 (2011)
6. J. Cashin, B. Lawson, Managing cyber supply chain risk - best practices for small entities, American Public Power Association, Washington, DC, Tech. Rep. (2018)
7. R. Spalding, Vulnerable 5G networks threaten world's critical infrastructure, Asia Times, Tech. Rep. (2021). [Online]. Available: https://asiatimes.com/2021/12/vulnerable-5g-networks-threaten-worlds-critical-infrastructure/
8. J. Kisielius, Breaking down the SolarWinds supply chain attack, SpyCloud, Tech. Rep. (2021). [Online]. Available: https://spycloud.com/solarwinds-attack-breakdown/
9. G. Gereffi, What does the covid-19 pandemic teach us about global value chains? the case of medical supplies. J. Int. Business Policy **3**(3), 287–301 (2020)

10. B.K. Mohanta, D. Jena, S.S. Panda, S. Sobhanayak, Blockchain technology: A survey on applications and security privacy challenges. Internet Things **8**, 100107 (2019)
11. S.A. Abeyratne, R.P. Monfared, Blockchain ready manufacturing supply chain using distributed ledger. Int. J. Res. Eng. Technol. **5**(9), 1–10 (2016)
12. K. Korpela, J. Hallikas, T. Dahlberg, Digital supply chain transformation toward blockchain integration, in *Proceedings of the 50th Hawaii International Conference on System Sciences* (2017)
13. D.J. Trump, Presidential executive order on assessing and strengthening the manufacturing and defense industrial base and supply chain resiliency of the united states (2017)
14. J. Villasenor, *Compromised by Design?: Securing the Defense Electronics Supply Chain* (Center for Technology Innovation at Brookings, 2013)
15. M. Willett, Lessons of the solarwinds hack. Survival **63**(2), 7–26 (2021)
16. N. Manworren, J. Letwat, O. Daily, Why you should care about the target data breach. Business Horizons **59**(3), 257–266 (2016)
17. J. Tidy, Honda's global operations hit by cyber-attack (2020). [Online]. Available: https://www. bbc.com/news/technology-52982427
18. J. Boyens, C. Paulsen, R. Moorthy, N. Bartol, Supply chain risk management practices for federal information systems and organizations. National Institute of Standards and Technology, Gaithersburg, MD, Tech. Rep. (2015)
19. M.J. Farooq, Q. Zhu, Modeling, analysis, and mitigation of dynamic botnet formation in wireless IoT networks. IEEE Trans. Inf. Forens. Secur. **14**(9), 2412–2426 (2019)
20. Q. Zhu, Z. Xu, *Cross-layer Design for Secure and Resilient Cyber-physical Systems* (Springer, 2020)
21. Q. Zhu, S. Rass, B. Dieber, V.M. Vilches et al., Cybersecurity in robotics: Challenges, quantitative modeling, and practice. Found. Trends® Robot. **9**(1), 1–129 (2021)
22. Q. Zhu, Control challenges, in *Resilient Control Architectures and Power Systems* (2021)
23. M.J. Farooq, Q. Zhu, On the secure and reconfigurable multi-layer network design for critical information dissemination in the Internet of battlefield things (IoBT). IEEE Trans. Wirel. Commun. **17**(4), 2618–2632 (2018)
24. Q. Zhu, S. Rass, On multi-phase and multi-stage game-theoretic modeling of advanced persistent threats. IEEE Access **6**, 13958–13971 (2018)
25. S. Rass, A. Alshawish, M.A. Abid, S. Schauer, Q. Zhu, H. De Meer, Physical intrusion games–optimizing surveillance by simulation and game theory. IEEE Access **5**, 8394–8407 (2017)
26. C. Folk, D.C. Hurley, W.K. Kaplow, J.F.X. Payne, The security implications of the Internet of things, AFCEA International Cyber Committee, Gaithersburg, MD, Tech. Rep. (2015)
27. T. Omitola, G. Wills, Towards mapping the security challenges of the Internet of things (IoT) supply chain. Procedia Comput. Sci. **126**, 441–450, 2018. [Online]. Available: http://www. sciencedirect.com/science/article/pii/S1877050918312547
28. K. Boeckl, M. Fagan, W. Fisher, N. Lefkovitz, K.N. Megas, E. Nadeau, B. Piccarreta, D.G. O'Rourke, K. Scarfone, Considerations for managing Internet of things (IoT) cybersecurity and privacy risks, National Institute of Standards and Technology, Gaithersburg, MD, Tech. Rep. (2019)
29. R.E. Hiromoto, M. Haney, A. Vakanski, A secure architecture for IoT with supply chain risk management, in *2017 9th IEEE International Conference on Intelligent Data Acquisition and Advanced Computing Systems: Technology and Applications (IDAACS)*, vol. 1 (2017), pp. 431– 435
30. Strategic principles for securing the Internet of things, U.S. Department of Homeland Security, Gaithersburg, MD, Tech. Rep. 2016. [Online]. Available: https://www.dhs.gov/sites/default/ files/publications/Strategic_Principles_for_Securing_the_Internet_of_Things-2016-1115- FINAL_v2-dg11.pdf
31. N. Bartol, Cyber supply chain risk management for utilities - roadmap for implementation, Utilities Telecom Council, Washington, DC, Tech. Rep. (2015)

Chapter 2
Risk Modeling and Analysis

Abstract Securing the supply chain of information and communications technology (ICT) has recently emerged as a critical concern for national security and integrity. With the proliferation of Internet of Things (IoT) devices and their increasing role in controlling real world infrastructure, there is a need to analyze risks in networked systems beyond established security analyses. Existing methods in literature typically leverage attack and fault trees to analyze malicious activity and its impact. In this chapter, we develop a security risk assessment framework borrowing from system reliability theory to incorporate the supply chain. We also analyze the impact of grouping within suppliers that may pose hidden risks to the systems from malicious supply chain actors. The results show that the proposed analysis is able to reveal hidden threats posed to the IoT ecosystem from potential supplier collusion.

2.1 Risk Scoring in Component Graphs

2.1.1 Introduction

Securing the supply chain and mitigating associated risk is critical to nearly every industry and enterprise. The Internet of Things (IoT) is capable of assisting supply chain risk management across different industries by enhancing tracking and monitoring capabilities [1]. However, the supply chain of the IoT itself faces risks that are much more complex and difficult to assess. The IoT is an integration of diverse systems involving informational technology (IT) and operational technology (OT) systems that are manufactured, owned, and operated by various different entities around the world. This opens doors for malicious actors to manipulate IoT systems and sabotage their operation [2]. Hence, maintaining supply chain integrity is now becoming a key concern for information and communications technology (ICT) users including governments and corporations [3].

Analyzing supply chain risks in IoT systems and networks is a crucial step in measuring the potential threat to the system and is a necessary precursor to making risk minimizing decisions regarding the supply chain. Supply chain risk management (SCRM) is a well studied subject in literature [4]. However, supply chain risks in IoT systems are fundamentally different from traditional industries such as food and medicine [5]. Among other reasons, the opacity of black box systems renders assessment of risks considerably more difficult. Furthermore, IoT supply chain risks are also different than the risks in traditional ICT systems due to their inter-connectivity. Suppliers themselves are inter-connected in ways that potentially increase risk, and these relationships are continually changing as firms adapt to market conditions. Hence, analyzing supply chain risks in IoT systems requires an adaptation of risk analysis techniques to consider these varied and complex sources of risk.

In this chapter, we propose a framework for IoT supply chain risk analysis and mitigation, referred to as I-SCRAM, that is centered around system components and their suppliers. This involves a shift from the perspective of traditional security risk analysis, where security events associated with particular component functions are primary. The core challenge involved in supply chain threats is that suppliers may potentially alter the system's functions in indeterminate ways. Therefore, any component could be a potential vector for such a threat, not only those marked as important for security. Risk analysis can take this into account only by widening the class of components under question and including their suppliers. This broader approach requires supply chain risk analysis to leverage tools from system reliability theory.

An example of a supply-chain attack would be when a supplier provides a product with a degraded implementation of a common cryptographic protocol, all the while claiming the product correctly uses the protocol to ensure the confidentiality of customer information. In the absence of extensive testing, the users of this product may assess the risk of using the product based on the (mis)information provided by the supplier. Not only does the product leave the customers open to attacks against the confidentiality of their data, the assessed risks are based on false information. Any mitigation efforts taken by the customer are significantly sub-optimal because of this supplier deception. Consequently, risk analysis must go beyond inherent security risks and take into account the trustworthiness of suppliers. Figure 2.1 illustrates the involvement of suppliers in the risk analysis process.

2.1.2   Related Work

Supply chain security of the IoT is an emerging area of research and is gaining considerable interest by the industry, government, and the news media. There have been some attempts in the literature to develop secure architectures for IoT with SCRM [6]. However, substantial analytical treatment of risk analysis and mitigation strategies in such complex system-of-systems scenario is still lacking. Existing literature related to SCRM falls broadly under three main categories:

Fig. 2.1 Supplier's role in security risk assessment. When providing a product, the supplier also provides associated specifications. These specifications form the basis of risk assessment of a component, but their accuracy affects the accuracy of the assessed risk

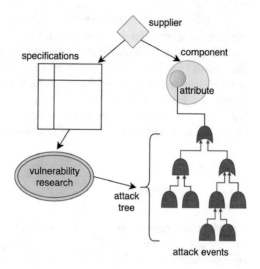

- Supply chain risk
- Graph-based security modeling
- Reliability analysis

Much interest is given to ensuring business continuity in the face of various disruptions to supply chains, for example as summarized in [4], though relatively little attention has been given to the security implications of supplier access to ICT products. A broad overview of cybersecurity risk posed by supply chains in a defense and national security context is presented in [7]. The U.S. National Institute of Standards and Technology (NIST) has issued a Special Publication on the topic of supply chain risk management that we use as a reference for terminology and associated risk management practices [8]. Best practices for SCRM in ICT systems have been published by NIST in [9]. Similarly, some strategic principles for securing the IoT are provided by the U.S. Department of Homeland Security in [10]. Our approach differs by providing a holistic modeling framework that supports systemic risk analysis as well as mitigation decisions.

A second area of related literature touches on security risk modeling techniques, in particular, those that use directed acyclic graphs. Helpful overviews of this large body of work are found in [11, 12]. The risk modeling approaches have been recently successfully applied to model the interdependencies among infrastructures subject to natural disasters [13–15] or human-induced attacks [16–18]. Attack trees, developed by Schneier and Amoroso, are adaptations from fault tree analysis in reliability theory [19, 20]. Significant developments have been made in the construction and use of attack graphs, including the incorporation of defensive measures, aggregation, and probabilistic modeling [12, 21–29]. A more formal analysis of attack trees is presented in [30, 31]. Our approach also employs a graph-based model. However, we have found it necessary to focus more on components and their suppliers rather than goals or security events, as is typically the case with attack graphs.

The third area of related literature is the extensive field of reliability theory, which provides many of the needed analytic tools for our approach, for example in [32, 33]. Baiardi et al. study hierarchical, hypergraph based modeling of systemic security risks across various security attributes [34]. This combination of system modeling and security attributes is a feature we find helpful for our analysis. Furthermore, significant research has also been conducted on reliability optimization [35]. However, we consider an optimization problem at the supply chain frontier, which takes into account the underlying system interconnections and the resulting risk.

2.1.3 Contributions

Security risk analysis has traditionally been studied in the context of estimating the probability of various, specific attack events against a system, a method incapable of capturing the possibility of covert modifications to the system itself. The security of supply chains has also been well studied in the literature using tools from graph theory and attack graphs. However, the interconnection of systems in an IoT setting results in a system-of-systems scenario that makes the analysis and mitigation of supply chain risk extremely challenging. In this work, we present a method to analyze system security by considering supplier trust and the relationships among suppliers. Our analysis is then used as a basis to solve risk mitigating decision problems involving the selection of suppliers. A summary of the main contributions of this work is as follows:

- We propose a system model, based on components and suppliers, for a unified analysis of traditional security risks and those arising from a supply chain.
- We study the quantification of supplier involvement in a system for individual suppliers as well as supplier groups.
- We develop a method to approach large scale supply chain risk mitigation decisions using the Birnbaum structural importance measure.
- We solve a risk minimizing supplier choice problem that considers supplier involvement to mitigate risks that may emanate from exclusive dominance of suppliers in the supply chain.

The contents of this chapter are organized as follows: First, a system model is defined in Sect. 2.2, that is capable of unifying supply chain risks along with traditional security risks. Then, we propose metrics for risk analysis in Sect. 2.3 using this system model. We then define the supplier choice problem as a risk mitigation decision, and provide a solution method in Sect. 3.2. Finally, we illustrate the use of I-SCRAM by a case study involving risk analysis and a supplier choice problem in Sect. 3.3 before concluding the chapter in Sect. 3.4.

2.2 System Model for Risk Assessment

We consider an abstraction of an IoT system involving a network of components and their individual suppliers. Our model takes into account a hierarchy of suppliers of the components and the logical network that defines the inter-dependencies of components and functionality of the system. We discuss each of these, in turn, before defining their relationship in terms of supplier trust and then presenting a unified system model.

2.2.1 Model Definitions

In this subsection, we describe the core elements of our system model that support risk analysis and mitigation decisions. These include the component security graph and the system supplier network, which are elaborated as follows.

2.2.1.1 Component Security Graph

The components in a system are related not only by functional dependencies, but also by security dependencies. Although the aim of I-SCRAM is to analyze the effect of suppliers on system security, we build this analysis on an understanding of the security dependencies among components themselves. The following are some of the key elements of the component security graph :

- *Components*, denoted by C, are the set of atomic system elements. The set of components in a system is retrieved through a process of hierarchical decomposition.[1] This can be carried out recursively to an arbitrary level of depth. For the purpose of supply chain risk modeling, a general rule is that a component should be a system element with a single or principal supplier. Any further decomposition would be unnecessary. Components may be hardware or software elements in general.
- *Component security attributes*, denoted by the set A, are Boolean properties of the component state relevant to security. The function determining the attribute value given the component state is defined in a security policy. Following convention, we use the following set of security attributes: {*confidentiality, integrity, availability*}. In principle, this set can be expanded based on the particular application at hand. For simplicity, we consider these attributes to have binary values: either the component state conforms to the security policy or not.

---

[1] Hierarchical decomposition refers to a process that takes a component in a system and considers it as a system in itself, returning subsystems and additional components.

- *Component security dependencies*, denoted by D are attributed to each component security attribute $n \in \mathbb{Z}^+$. Each dependency is a component, attribute tuple: $\langle c \in C, a \in A \rangle$. The security failure of a dependency affects the component in question. For an attribute to be *true*, a necessary but not sufficient condition is that its dependencies must evaluate to *true*. A security dependency captures a relationship between components similar to edges in an attack graph. However, component security dependencies do not consist of events as in an attack graph. Each dependency rests in a component, which is the core abstraction in our model rather than events. Each component, attribute tuple may be considered a potential *root* node for an attack tree, in which at least some *attack events* coincide with other component, attribute tuples.

- The *component security logic function*, ℓ, computes the value of the vector of component security dependencies. We restrict our analysis to the use of two logic functions: $\{AND, OR\}$. These logic functions encode whether or not the state of a component's security dependencies is sufficient to cause a failure in the component's security attribute. For example, if the component, attribute tuple $\langle j, c \rangle$ possesses dependencies $\{\langle p, c \rangle, \langle q, c \rangle\}$, with the logic function AND, then a security failure in either $\langle p, c \rangle$ or $\langle q, c \rangle$ yields a dependency state of *false*, causing a security failure in $\langle j, c \rangle$.

- *Component security risk*, s, is the probability that $\langle c, a \rangle = false$ and is computed as the union of $\ell(D)$ and a risk value local to the component attribute itself, referred to as r. Security failures can happen either by a node directly being attacked, or by a sufficient attack against its dependencies. Given both probabilities, the risk at any node can be calculated. However, care must be taken in computing $\ell(D)$ because dependencies may not be independent.[2] This difficulty can be eliminated in various ways, for example, by restricting valid system configurations to a tree or by taking the dependence into account. Because such approaches reduce the model's utility, in particular when adding the complexities of supply chain topologies, our approach relies on the use of minimal cutsets as explored in reliability theory [36].

In light of the above elements, a component security graph is formally described by Definition 2.1.

Definition 2.1 A *component security graph* is defined as a connected, directed graph:

$$G_c = (V, E), \text{ where} \tag{2.1}$$

---

[2] In other words, given two dependencies $a, b \in D$, there may be some node x such that $x \in D_a$ and $x \in D_b$. In such a case it would be invalid to compute $\ell(D)$ simply from the suppliers of a and b, denoted by s_a and s_b respectively, because s_a and s_b are not independent.

$$V = \{\langle c, a \rangle \mid c \in C, a \in A\}, \text{ and}$$
$$E = \{\langle d, v \rangle \mid d \in D_v, v \in V\}.$$

When considering general system security risk, it is necessary to specify what constitutes system-wide security. Analogous to Leveson's discussion of system safety in [37], security can be considered as an emergent property in a complex system. As such, it is a non-trivial function of the component security graph state. Our approach relies on a subset of components chosen as indicator nodes aggregated by a logic function that, if *false*, represents a system security failure. Alternate modeling approaches may be more suited to more diffuse threat scenarios, such as a summation over expected loss values.

2.2.1.2   System Supplier Network

In this section we consider definitions relevant to the supply chain portion of the problem. The key elements of the supplier network are described as follows:

- *Suppliers*, denoted by S, are any "organization or individual that enters into an agreement with the acquirer or integrator for the supply of a product or service" [8]. Every $s \in S$ possesses at least one component π_s as its corresponding product. Inversely, s_c refers to the supplier of c. Service providers correspond to the product for which they provide service. We consider the category of suppliers as widely as possible, including for example design, manufacturing, logistics, retail, maintenance, and disposal. In every case, the entity in question has significant opportunities for extended, intrusive, and covert access to the component. This access is considered essential to being a supplier of the component. Consideration of differing degrees of access may be incorporated into supplier trust, considered below.
- A *supplier group* G_k consists of a finite set of mutually involved suppliers capable of coordinated action.[3] It is not required that every member of a group have access to the products of the other members as if they were their own product. Rather, it suffices that the group's relationship supports coordinated action. Each group G_k possesses a *supplier group controller*, κ_k. This entity directs any coordinated action by the group. If $\kappa_k \in G_k$, the group is directed by a member. Otherwise, a non-member directs group action. If the group possesses a decentralized organization structure, for simplicity we abstract this by the creation of a fictitious controller that represents group action. No group controller possesses any relationship to a component product except through the members

---

[3] The precise legal relationships that may constitute a supplier group are left unspecified here, but may include ownership, partnership, or membership in joint ventures or cartels whether legally recognized or not.

of the group it directs. In other words, controllers are not suppliers in the strict sense.

- The *universal supplier network* denotes the complete set of suppliers across an industry, the suppliers of components acquired by them, and the interconnections between these entities. The network is defined as a directed graph:

$$G_u = (V, E), \tag{2.2}$$

where

$$V = \{s \mid s \in S\}, \text{ and } E = \{\langle s, s' \rangle \mid \langle s, s' \rangle \in S \times S\}.$$

It is assumed that all products are composed of components either purchased or manufactured, entailing that the global supplier network includes each supplier of each component at each level of complexity across every technical system. The scope of this network is intentionally general, as it functions like the universal set against which to define more specific networks. Edges in the supplier network constitute a supply chain. The precise significance of an edge $\langle s, s' \rangle$ is that in the case of some product, supplier s' functioned as an 'acquirer or integrator' as referenced in the definition of supplier, with s as supplier. A *particular supplier network* is an induced subgraph, $G_u[S'_c]$, where S'_c is defined for some component c as:

$$S'_c = \{s \mid s = s_c\} \cup \{s' \mid \text{a path exists from } s' \text{ to}$$

$$s \in G_u\}. \tag{2.3}$$

Based on the above definitions, the system supplier network is formally described by Definition 2.2.

Definition 2.2 The *system supplier network*, denoted by G_s, defines the set of suppliers with connections to some component security graph G_c, along with their dependencies. It is defined as:

$$G_s = \bigcup_{c \in G_c} G_u[S'_c]. \tag{2.4}$$

It should be noted that when defining supplier networks, no reference is made to security attributes. Only components and their suppliers are considered. The source of supply chain risk that we seek to model is the particular degree of contact between a supplier and a component, a property that is much more general than any security attribute. Indeed, because a supplier may potentially manipulate a component in relatively unlimited ways, any security attribute possessed by the component may be affected by its supplier. For example, the supplier of the plastic chassis of a device *should not* have relevance for the device's *confidentiality* attribute. However, because this supplier has extensive access to the device there may be some risk that it has tampered with the device. For example, perhaps it has inserted an eavesdropping device. This kind of modification vastly exceeds what is expected of a supplier but is neither inconceivable nor insignificant to risk modeling.

2.2.2 Supplier Trust

In this section we present definitions relevant to supplier trust, a central part of our model. For each supplier and supplier group in the system under analysis, a value in the interval [0,1] indicates the extent to which the entity is trusted. Although trust has a familiar and intuitive meaning, we find it necessary to offer a formal definition in order to justifiably relate supplier trust to security risk. Every system is an implementation of a set of functions, achieved by the coordination of the functions of individual system components. The question of supplier trust is: which functions does this system implement, how do we know, and how confident are we in this knowledge?

Actual functions, denoted by A, are those functions that a component implements successfully within expected reliability standards. When provided with proper input, the component reliably produces the proper output. In practice, however, ascertaining the membership of set A for any arbitrary component may not be straightforward. At best, a process of empirical study can be conducted, but epistemological limitations make an exhaustive enumeration of the actual functions of a component impossible. This limitation is relevant to the issue of supplier trust. While verification of the presence of a function can be conducted by testing, it is not possible strictly to ascertain with certainty all the functions that a component *does not* implement. This points to the need for a well-grounded quantification of trust.

We define *putative functions*, P, as the set of functions of a component that are allegedly implemented, according to its supplier. These putative functions may equally well be called technical specifications. When a supplier provides a component, let us assume the supplier also provides a set of specifications that indicate the functions implemented by the component. It is generally assumed that relevant details are not accidentally omitted. If some omission in fact occurs, we do not distinguish here between potential *sources of fault* for the omission.

Hidden functions belong to the set $A \setminus P$. They are implemented in fact without being communicated to acquirers. They may be benign or malicious. On the contrary, *Missing functions* belong to the set $P \setminus A$. Though the supplier claims they have been implemented, in fact they are not. With the above definitions in mind, supplier trust is formally expressed by Definition 2.3.

Definition 2.3 *Supplier trust* is related to the set of putative functions, P and the set of actual functions, A, using the Jaccard index [38]:

Fig. 2.2 Supplier trust
involves the overlap of
putative and actual functions
of a component

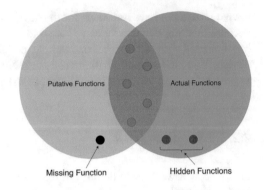

$$t_s = \frac{|P \cap A|}{|P \cup A|}. \tag{2.5}$$

We employ the notation $\bar{t}_s = 1 - t_s$ for supplier risk.

The trustworthiness of a supplier indicates the degree to which the putative functions of a component and its actual functions overlap. This is illustrated in Fig. 2.2, where both missing and hidden functions negatively impact trust. Trust can likewise be interpreted as a measure of the strength of belief that stated specifications are accurate and exhaustive.

Based on supplier trust, we define the *adjusted component security risk* as:

$$\hat{r} = \bar{t}_s + r_c - \bar{t}_s r_c \text{ where } s = s_c. \tag{2.6}$$

Following the above discussion of supplier trust, risk arises in a component from either the component itself or the inaccuracy of the supplier's provided specifications. The usefulness of our particular definition of trust lies in its basis in component functionality. Component security risk is estimated by experts with considerable reliance on the putative functions of a component. The degree of belief in the accuracy of this information impacts the confidence one can reasonably have in the resulting risk value. The supplier therefore stands in a critical though often overlooked relationship to the process of component risk assessment. This relationship is illustrated in Fig. 2.1. It is pertinent to mention that in the case of a hierarchy of suppliers, the component risk will be impacted by a compounded trust of individual suppliers in the chain.

We define the *supplier trust estimation problem* as assigning t_s for some supplier s. There are two methods to approach the determination of supplier trust values. The first method is deductive and descriptive, closely following the definition above. Supposing an exhaustive or reasonably exhaustive knowledge of A for the components provided by a supplier, the value of t_s is computed following Definition 2.3. The value t_s describes the trustworthiness of the supplier based on objective criteria. The second method is inductive and useful for predictions. By assigning a trust value through a heuristic process, the value can be used to predict

Fig. 2.3 A simple systemic risk graph composed of components and suppliers with security logic functions indicated

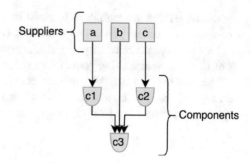

the expected degree of overlap between P and A, avoiding the issue of how to ascertain the membership of A. We pass over the description of such a heuristic process, while referencing existing practices in security risk assessment techniques [8, 39].

2.2.3 Systemic Risk Graph

The above definitions form the basis of the unified system model that is used for risk analysis and mitigation decisions.

Definition 2.4 Gathering the above definitions, a *systemic risk graph*, G_r, is defined as the composition of a component security graph and the relevant system supplier network, as follows:

$$G_r = G_c \oplus G_s \tag{2.7}$$

where G_s is defined from G_c and '$\oplus$' represents the graph composition operator.

As an example, a simple system with three components is shown in Fig. 2.3. In this example, keeping with the simplified model presented here, the possible existence of multiple, relevant security attributes at a component is left aside. Component c_3 is marked as an *AND* node, with two components c_1, c_2 indicated as dependencies by directed edges. Because c_3 is *AND*, the security of c_3 depends on not only its own security but also both c_1 and c_2, as well as their suppliers a, b and c.

2.3 Risk Analysis Metrics

At this point all the elements of the I-SCRAM system model have been discussed. Suppliers and components are related in a systemic risk graph that can be then used to compute various measures of risk. In the following definitions, several measures are discussed that are then used to assist in system optimization. Given the above definitions, the security risk in a system may be approached generically as a problem analogous to system reliability. The risk analysis conducted here follows the method of distilling a systemic risk graph into a set of minimal cutsets [32].

The *minimal cutsets*, W, indicate possible system states in which a critical security failure, predefined with an indicator node i, has occurred. Each cutset $w \in W$ is a subset of $\{v \mid v \in G_r\}$ such that $i = 0$ if

$$1 - \prod_{v \in w}(1 - v) = 0.$$

A cutset w is minimal when it cannot be reduced without failing to be a cutset, i.e., when $\nexists z \subset w$, such that $z \in W$. As noted above, a target that represents the system security state must be specified. For example, three significant component, attribute tuples might be chosen and the system security indicator might be defined as an AND node with these as dependents. A minimal cutset then represents a way to cut the indicator from its dependencies. Note that nodes can fail on their own even if their dependents are functioning. Procedures for discovering the minimal cutsets in a system graph are borrowed from reliability theory. We note here the computational complexity of the discovery of minimal cutsets. The MOCUS algorithm, which is capable of handling nodes with AND and OR logic, can improve on the worst case in practice [40]. Considerable research has been conducted on efficient implementations of this and other algorithms relevant to fault tree analysis [41, 42].

2.3.1 Systemic Risk Function

The general *systemic risk function* computes the probability that all of the nodes have failed of at least one of the minimal cutsets. Given a vector of risk values $\mathbf{r}$, it is defined as follows:

$$R(\mathbf{r}) = 1 - \prod_{w \in W}\left(1 - \prod_{v \in w} r_v\right) \tag{2.8}$$

The *Birnbaum structural importance measure*, I_i, measures the sensitivity of system risk to the risk of some component i. It is closely related to the *Birnbaum component importance measure*, which is a partial derivative of system risk with respect to individual component risk:

$$I_i = \frac{\partial R(\mathbf{r})}{\partial r_i}$$

System reliability analysis has produced several measures of component importance, the most relevant of which is the Birnbaum structural importance measure [32]. Although the measure is typically used with reference to reliability, we employ it to consider risk. It has been shown that the two measures are equal for component i when the risk of every component $j \neq i$ is 0.5. While referring to established sources for more ample discussion of the calculation of this value, we note the following. Let two vectors be defined as:

$$s_i^1 = \{r_i = 1.0\} \cup \{r_j = 0.5 \mid j \in \{1, \ldots, n, j \neq i\}\}$$
$$s_i^0 = \{r_i = 0.0\} \cup \{r_j = 0.5 \mid j \in \{1, \ldots, n, j \neq i\}\}$$

The Birnbaum structural importance measure can then be computed as:

$$I_i = R(s_i^1) - R(s_i^0)$$

We define the *risk importance measure, RI* as the inherent security risk of the component weighted by the Birnbaum structural importance measure of the component. It is defined as follows,

$$RI_c = r_c \times I_c \tag{2.9}$$

2.3.2 Supplier Involvement Measure

A *supplier involvement measure, SI*, captures the degree to which any supplier or supplier group is involved in a systemic risk graph. This measure includes consideration of supplier trust, component importance, as well as the grouping of suppliers. We quantify the supplier involvement measure as follows:

$$SI_k = \left(\sum_{j \in G_k} I_{\pi_j} \right)^2 \bar{t}_k \tag{2.10}$$

Several aspects of this measure must be noted. First, the involvement of each supplier j is captured by I_{π_j}, the Birnbaum structural importance of the component it supplies. Secondly, the aggregate involvement of a supplier group is squared. A critical aspect of supply chain security consists of the risk involved when components are supplied by entities that are distinct yet organized into a supplier group capable of coordinated action. We believe that the supplier involvement measure should reflect this risk by being amplified as group size increases. Finally, the value $\bar{t}_k$ is the risk value corresponding to the supplier group controller. As a technical note regarding implementation, a supplier j that does not belong to any supplier group k can be considered in its own group with $\bar{t}_k = 0.0$.

2.4 Uncertainties in Model Development

Although the model we have developed bears a resemblance to established methods of system reliability analysis, the problem domains of reliability and security differ sufficiently to warrant a discussion of limits and challenges to the use of this model in practice. While certain of these challenges may be overcome through developments in methodology, others may point to limits within which the problem of supply chain security analysis must be conducted. In the analysis that follows we identify major challenges to the accurate construction of this model for a real system. After discussing these challenges in general, we illustrate the effect of four kinds of uncertainties using a case study.

2.4.1 Parametric Uncertainties in Probability Estimates

The first major area of difficulty in the use of such a model is obtaining accurate probability estimates for basic events such as component security failures and, more critically, supplier security failures. Estimating the likelihood of a supplier being compromised or being covertly malicious is a problem involving considerable difficulty. On the assumption that any compromise or malicious act will eventually be detected and attributed accurately, the accuracy of risk values will generally increase as this information is incorporated into assessed likelihoods. We consider the problem of estimating accurate risk values to be best approached through the development and use of heuristics and metrics together with information gathering and regular assessments. If accuracy is a limitation here then it is one that system design and use must accommodate.

2.4.2 Structural Modeling Uncertainties

A second source of uncertainty lies in the possibility that sources of risk are simply omitted from the system model, i.e., that some set of nodes or edges that *should* be in the system graph are not included. We call these structural uncertainties, and define them as a modeling choice that has some effect on the set of minimal cutsets. Therefore the three kinds of uncertainties here will include those related to nodes, edges, and node logic functions. Being uncertain about the structure of a system could easily be a matter of neglect or oversight, but may just as well be a result of a complexity in system design that lies outside the reasonable purview of those building the model. In the case of both an inaccurate probability value and a structural modeling divergence, the calculated systemic risk value does not correspond to the real system risk. In the case studies that follow, we seek to illustrate the observation that structural uncertainties pose a significant challenge to accurate modeling and merit priority over improvements in accurate probability estimations.

2.5   Uncertainty Case Studies

In the case studies that follow, we first present a ground truth scenario that is intended to represent an ideal system graph constructed to model the system in question. Following this, we discuss four kinds of uncertainties and illustrate the possible effect of each by comparing the results of risk analysis after each error with the ground truth scenario.

2.5.1   Case 0: Ground Truth

The system graph for this case study is shown in Fig. 2.4. It possesses a tree structure with twenty-five nodes and with roughly equal numbers of AND and OR nodes distributed throughout the graph. We have chosen the tree structure to provide the basis of these examples because of its resemblance to classical fault trees. In practice the structure could vary widely. However, for the purpose of this study a tree structure seems likely to provide a suitable basis for generalization.

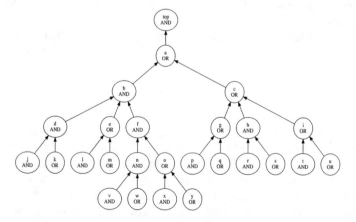

Fig. 2.4 System Graph for Case 0: Uncertainty scenarios will be examined with reference to this as the ground truth scenario. System security is represented by the top node, and node failures that constitute minimal cutsets will cause a failure of the top node

Table 2.1 Minimal cutsets for case 0: Each cell contains one minimal cutset such that a failure of every node in the cutset entails a system security failure

{a}	{b}	{c}
{g}	{h}	{i}
{p}	{q}	{t}
{u}	{r,s}	{d,e,f}
{d,f,l}	{d,f,m}	{d,e,n,o}
{d,e,n,x}	{d,e,n,y}	{d,l,n,o}
{d,l,n,x}	{d,l,n,y}	{d,m,n,o}
{d,m,n,x}	{d,m,n,y}	{e,f,j,k}
{f,j,k,l}	{f,j,k,m}	{d,e,o,v,w}
{d,e,v,w,x}	{d,e,v,w,y}	{d,l,o,v,w}
{d,l,v,w,x}	{d,l,v,w,y}	{d,m,o,v,w}
{d,m,v,w,x}	{d,m,v,w,y}	{e,j,k,n,o}
{e,j,k,n,x}	{e,j,k,n,y}	{j,k,l,n,o}
{j,k,l,n,x}	{j,k,l,n,y}	{j,k,m,n,o}
{j,k,m,n,x}	{j,k,m,n,y}	{e,j,k,o,v,w}
{e,j,k,v,w,x}	{e,j,k,v,w,y}	{j,k,l,o,v,w}
{j,k,l,v,w,x}	{j,k,l,v,w,y}	{j,k,m,o,v,w}
{j,k,m,v,w,x}	{j,k,m,v,w,y}	

Table 2.2 Results for case 0, ground truth

$\|W\|$	53
$avg(\|w\|) \; \forall w \in W$	4.018868
$J(W, W')$	0.0
$Risk$	0.403032
$\Delta Risk$	0

The minimal cutsets of this system are shown in Table 2.1. Each cell contains a set of nodes identified by alphabet letter, where the security failure of all nodes in the set represents a security failure of the top node. To compute a risk value for the system, we provide sample component risk values such that each component has a risk of failure of 0.05. Essential metrics for this system are found in Table 2.2.

2.5.2   Case 1: Uncertainty of Single Node Logic

In this first uncertainty scenario, the logic type of a single node will be modified to represent the mis-classification of a node with regard to its predecessors. A full treatment of the case of a single logic error suggests investigating the effect of this error on any given node in the graph. While this would indeed yield a more thorough understanding, such a generalized study would be of limited value without operating on a generalized graph. In lieu of this theoretical exercise, here we present the result of analysis when various nodes in the case study are mistaken. We chose nodes c and b, where the analysis will be conducted on the graph after each node's logic function $\ell_n \in \{AND, OR\}$ has been substituted for the opposite type. Descriptively, this

Table 2.3 Results for case 1,
logic uncertainty in c

$|W|$	63
$avg(|w|) \; \forall w \in W$	4.238095
$J(W, W')$	0.366197
$Risk$	0.144027
$\Delta Risk$	−0.259005

Table 2.4 Results for case 1,
logic uncertainty in b

$|W|$	23
$avg(|w|) \; \forall w \in W$	1.478261
$J(W, W')$	0.830769
$Risk$	0.542643
$\Delta Risk$	0.139611

entails an error in recognizing the way that components g, h, i and their predecessors affect the security of component c, with an analogous error for node b. While the ground truth scenario includes a more risk-amplifying relationship, where any of g, h, i can cause c to fail, the situation studied in Case 1 is that the model designer considers c to fail when all of g, h, i have failed. As such, we expect that this analysis will result in an erroneously low risk assessment when node c has been modified, while the opposite will be the case for node b.

Detailed results for Case 1 on node c are shown in Table 2.3, including a modest rise in the number of cutsets as well as their average size. The Jaccard distance from the ground truth is 0.366, indicating a probability of roughly 1/3 that a cutset in either case is not shared between the two. Risk, as expected, has dropped by 0.259. To contrast, we present the results for changing node b in Table 2.4. This single node logic error results in a significant Jaccard distance of 0.83, while raising systemic risk by 0.139. Finally, Fig. 2.5 shows the effect of a logic error at each node in the system. Because many nodes are leaf nodes possessing no dependencies, the nature of the logic function at the node is irrelevant to systemic risk. Similarly, we note a general correlation between the magnitude of the change in risk and the height of the node in question.

2.5.3   Case 2: Uncertainty of Node Omission

When constructing a model, it may easily occur that a component is overlooked and omitted from the model. Especially as complex systems involve many layers of components, there will be some uncertainty concerning whether important nodes have been omitted. To capture this uncertainty, we test here the result of deleting a node from the system graph. As in Case 1, much of the effect of such an error will depend on the topography of the system graph as well as the location of the node omitted. When omitting a node, we consider it necessary to omit also the children of the node that become disconnected from the graph as a result. This

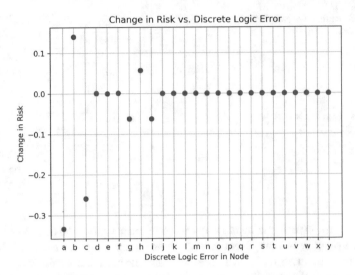

Fig. 2.5 Case 1 Results: When node logic is subject to discrete error, systemic risk values vary widely, but with a magnitude related to node height

choice reflects the likelihood that in overlooking a component, its sub-components or dependencies will also be overlooked. We test here the omission of a mid-level node f and, separately, a higher-level node c. Fig. 2.6 shows the modified system graph with node c omitted.

Detailed results are shown in Tables 2.5 for the omission of node f and 2.6 for the omission of node c, while a survey of the resulting change in risk for each node's omission is shown in Fig. 2.7. It is pertinent to note the lack of correlation between the Jaccard distance of the minimal cutsets and the change in systemic risk. While omitting node f yields a very significant distance between the cutsets (0.81), the change in risk is minimal (0.004). By contrast, omitting node c results in a smaller Jaccard distance (0.17) but a very large decrease in risk (0.305). This volatility in modeling results points toward the importance of component level analysis in understanding supply chain risk. We also note the lack of correlation here between the change in risk and the height of the node in question. Omitting node c has a rather large effect, whereas node b, with the same height, has a very small effect.

2.5.4   Case 3: Uncertainty in Edge Placement

The scenario captured as uncertainty in the placement of a single edge will be when a component node is successfully identified but it is mistaken how the node is related to other nodes in the system. As such, an edge error entails no change in the number of connected nodes in the graph. There may be a large number of possibilities that are plausible ways an edge might be mistakenly placed. As such it will be difficult

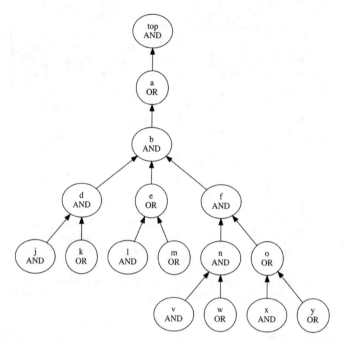

Fig. 2.6 System graph for case 2, illustrating an erroneous omission of the component at node *c* as well as its sub-components

Table 2.5 Results for case 2, node omission in *f*

$|W|$	17
$avg(|w|) \; \forall w \in W$	1.588235
$J(W, W')$	0.813559
Risk	0.407450
$\Delta Risk$	0.004418

Table 2.6 Results for case 2, node omission in *c*

$|W|$	44
$avg(|w|) \; \forall w \in W$	4.613636
$J(W, W')$	0.169811
Risk	0.097911
$\Delta Risk$	-0.305121

to examine all the nodes as was done in the previous two cases. We present detailed results of two different edge errors. First, we remove edge $\langle d, b \rangle$ and substitute it for the edge $\langle d, e \rangle$. The results of this error are shown in Table 2.7. To contrast, we also investigate the change of edge $\langle h, c \rangle$ to $\langle h, g \rangle$, the results of which are shown in Table 2.8.

In the first examined edge modification we find a minimal change in risk despite a large distance between the cutsets. In contrast, a change in the edge $\langle h, c \rangle$ yields the identical minimal cutsets, owing to the nature of the original parent node's logic.

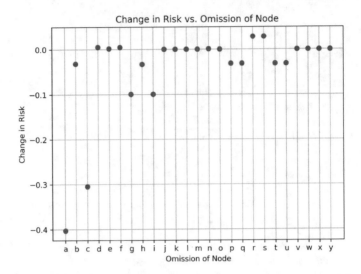

Fig. 2.7 Case 2 results, showing the change in system risk when a single node and the sub-tree rooted at the node is omitted. The magnitude of change in risk is not strictly correlated with node height

Table 2.7 Results for case 3, edge $\langle d, b \rangle \rightarrow \langle d, e \rangle$

$\lvert W \rvert$	46
$avg(\lvert w \rvert)\ \forall w \in W$	2.913043
$J(W, W')$	0.875
$Risk$	0.409726
$\Delta Risk$	0.006694

Table 2.8 Results for case 3, edge $\langle h, c \rangle \rightarrow \langle h, g \rangle$

$\lvert W \rvert$	53
$avg(\lvert w \rvert)\ \forall w \in W$	4.018868
$J(W, W')$	0.0
$Risk$	0.403032
$\Delta Risk$	0.0

2.5.5   Case 4: Uncertainty in Probability Values

After having explored the various kinds of structural uncertainties and illustrated their potential effects in particular cases, we examine here the contrasting effects of uncertainties in the estimation of probability values. These probability values are critical points of data without which a model cannot approximate the real risk in a system. Yet because of the difficulty of obtaining these values with accuracy and confidence, we examine the general effect of various margins of error. When applying each margin of error, $0 < e \leq 1$, the adjustment is made by adding er_i to element i of vector $\mathbf{r}$. With this adjusted vector, the general risk function is calculated. Because this class of errors involves no change to the number or identity

Table 2.9 Results for case 4

e	0.02	0.05	0.10	0.50
$Risk$	0.409364	0.418751	0.434108	0.544767
$\Delta Risk$	0.006332	0.015719	0.031076	0.141735

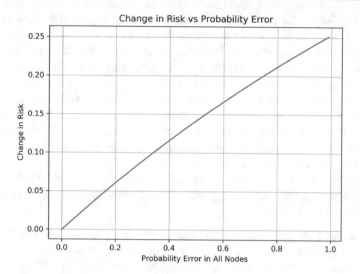

Fig. 2.8 Case 4 Results, where the probabilities of all events are adjusted by increasing margins of error. Elevated probability values entail a linear increase in risk, but only high margins of error are comparable to many structural uncertainties

of cutsets, we only compare the resulting risk value to the ground truth scenario presented above. In Table 2.9, we show the effect on risk analysis of four margins of error, e: 2%, 5%, 10% and 50%. Fig. 2.8 shows a range of errors and the resulting change in systemic risk.

We note that errors are calculated with reference to the ground truth scenario, where the probability of each event is 0.05. As such, the maximum error shown in Fig. 2.8, 100%, results in an adjusted probability of 0.10. Likewise, we apply this margin of error to every node in system graph. While more complex or drastic scenarios can be imagined, these high error rates are sufficient to illustrate the relative impact of uncertainties of different kinds.

2.6 Conclusion

In this chapter we have presented a modification of attack tree modeling suited for the analysis of supply chain risks with the primary intention of investigating the practical utility of such a model when faced with inevitable difficulties in obtaining accurate data describing complex ICT and IoT critical infrastructure systems. The preceding case studies have depicted various possible error scenarios that may be encountered while applying this modeling technique to an existing system. Using a particular system graph, we have illustrated these error scenarios with the help of several examples. Although caution is warranted when approaching a problem of this complexity from particular case studies, we use the results shown here to highlight the importance of structural errors in comparison to errors in obtaining accurate probability estimates. If the security of components and the trustworthiness of suppliers can be estimated to within 50% accuracy, our results show a maximum possible error in risk assessment of 14%. This is a significant change in risk, but equal or far greater discrepancies are found with a wide variety of discrete structural errors. Mistaking a single node's logic function, if it is a systemically important node, may produce double the change in assessed risk. Similar discrepancies are found with single node omissions or mistaken edges. Generally, the nodes with greater height in the system graph are more conducive to yielding larger discrepancies in systemic risk.

From these observations certain principles in the practical development of supply chain risk assessments may be suggested. The following preferences summarize the conclusions of this study.

- **Structure over magnitude:** Given the scarcity of resources available to conduct risk assessments, and the possible impact of errors of various kinds, we suggest significant attention be given to accurate structural modeling. While efforts to obtain accurate magnitudes in risk and trust values are certainly important, the development of accurate structural models for the ways in which components relate to each other as security dependencies should usually be prioritized.
- **Height over depth:** At higher levels of systemic analysis, accuracy in structural modeling should take unambiguous priority. Structural errors in the critical window of 2–3 hops from the top event have the potential to make extraordinary differences in modeling results. Extensive and accurate modeling of depth into a system may be helpful, but it is less important than ensuring accuracy in this critical window. The difficulty of obtaining accurate modeling at lower levels is matched by a decrease in the impact of possible errors. As such, less effort should be expended on components at these lower levels of a system.

References

1. C.K. Wu, K.F. Tsang, Y. Liu, H. Zhu, Y. Wei, H. Wang, T.T. Yu, Supply chain of things: A connected solution to enhance supply chain productivity. IEEE Communications Magazine 57(8), 78–83 (2019)
2. C. Folk, D.C. Hurley, W.K. Kaplow, J.F.X. Payne, The security implications of the Internet of things, AFCEA International Cyber Committee, Gaithersburg, MD, Tech. Rep. (2015)
3. A. Levite, ICT supply chain integrity: Principles for governmental and corporate policies (2019)
4. C.S. Tang, Perspectives in supply chain risk management. Int. J. Prod. Econ. **103**(2), 451–488 (2006)

5. T. Omitola, G. Wills, Towards mapping the security challenges of the Internet of things (IoT) supply chain. Procedia Comput. Sci. **126**, 441–450 (2018)
6. R.E. Hiromoto, M. Haney, A. Vakanski, A secure architecture for IoT with supply chain risk management, in *9th IEEE International Conference on Intelligent Data Acquisition and Advanced Computing Systems: Technology and Applications (IDAACS 2017)*, vol. 1 (2017), pp. 431–435
7. C. Nissen, J. Gronager, R. Metzger, H. Rishikof, Deliver uncompromised: A strategy for supply chain security and resilience in response to the changing character of war, Mitre Corporation, Tech. Rep. (2018)
8. J. Boyens, C. Paulsen, R. Moorthy, N. Bartol, Supply chain risk management practices for federal information systems and organizations, National Institute of Standards and Technology, Gaithersburg, MD, Tech. Rep. (2015)
9. K. Boeckl, M. Fagan, W. Fisher, N. Lefkovitz, K.N. Megas, E. Nadeau, B. Piccarreta, D.G. O'Rourke, K. Scarfone, Considerations for managing Internet of things (IoT) cybersecurity and privacy risks, National Institute of Standards and Technology, Gaithersburg, MD, Tech. Rep. (2019)
10. Strategic principles for securing the Internet of things, U.S. Department of Homeland Security, Gaithersburg, MD, Tech. Rep. (2016). [Online]. Available: https://www.dhs.gov/sites/default/files/publications/Strategic_Principles_for_Securing_the_Internet_of_Things-2016-1115-FINAL_v2-dg11.pdf
11. B. Kordy, L. Piètre-Cambacédès, P. Schweitzer, DAG-based attack and defense modeling: Don't miss the forest for the attack trees. Comput. Sci. Rev. **13**, 1–38 (2014)
12. W. Xiong, R. Lagerström, Threat modeling–a systematic literature review. Comput. Secur. **84**, 53 (2019)
13. R. Zimmerman, Q. Zhu, F. de Leon, Z. Guo, Conceptual modeling framework to integrate resilient and interdependent infrastructure in extreme weather. J. Infrastructure Syst. **23**(4), 04017034 (2017)
14. R. Zimmerman, Q. Zhu, C. Dimitri, Promoting resilience for food, energy, and water interdependencies. J. Environ. Stud. Sci. **6**(1), 50–61 (2016)
15. R. Zimmerman, Q. Zhu, C. Dimitri, A network framework for dynamic models of urban food, energy and water systems (fews). Environ. Prog. Sustain. Energy **37**(1), 122–131 (2018)
16. L. Huang, J. Chen, Q. Zhu, A large-scale Markov game approach to dynamic protection of interdependent infrastructure networks, in *International Conference on Decision and Game Theory for Security* (Springer, 2017), pp. 357–376
17. L. Huang, J. Chen, Q. Zhu, Distributed and optimal resilient planning of large-scale interdependent critical infrastructures, in *2018 Winter Simulation Conference (WSC)* (IEEE, 2018), pp. 1096–1107
18. L. Huang, J. Chen, Q. Zhu, A factored MDP approach to optimal mechanism design for resilient large-scale interdependent critical infrastructures, in *2017 Workshop on Modeling and Simulation of Cyber-Physical Energy Systems (MSCPES)* (IEEE, 2017), pp. 1–6
19. B. Schneier, Attack trees: A formal, methodical way of describing the security of systems, based on varying attacks. Dr. Dobb's J. **12**, 21 (1999)
20. E.G. Amoroso, *Fundamentals of Computer Security Technology* (PTR Prentice Hall, Englewood Cliffs, 1994)
21. A. Roy, D.S. Kim, and K.S. Trivedi, Attack countermeasure trees (ACT): towards unifying the constructs of attack and defense trees. Secur. Commun. Netw. **5**(8), 929–943 (2012)
22. J. Homer, S. Zhang, X. Ou, D. Schmidt, Y. Du, S.R. Rajagopalan, A. Singhal, Aggregating vulnerability metrics in enterprise networks using attack graphs. J. Comput. Secur. **21**(4), 561–597 (2013)
23. L. Wang, T. Islam, T. Long, A. Singhal, S. Jajodia, An attack graph-based probabilistic security metric, in *IFIP Annual Conference on Data and Applications Security and Privacy* (Springer, 2008), pp. 283–296
24. M. Gribaudo, M. Iacono, S. Marrone, Exploiting bayesian networks for the analysis of combined attack trees. Electron. Notes Theoret. Comput. Sci. **310**, 91–111 (2015)

25. N. Poolsappasit, R. Dewri, I. Ray, Dynamic security risk management using Bayesian attack graphs. IEEE Trans. Dependable Secure Comput. **9**(1), 61–74 (2011)
26. O. Sheyner, J. Haines, S. Jha, R. Lippmann, J.M. Wing, Automated generation and analysis of attack graphs, in *Proceedings IEEE Symposium on Security and Privacy* (2002), pp. 273–284
27. X. Ou, W.F. Boyer, M.A. McQueen, A scalable approach to attack graph generation, in *Proceedings of the 13th ACM Conference on Computer and Communications Security* (2006), pp. 336–345
28. Z. Qian, J. Fu, Q. Zhu, A receding-horizon MDP approach for performance evaluation of moving target defense in networks, in *2020 IEEE Conference on Control Technology and Applications (CCTA)* (IEEE, 2020), pp. 1–7
29. L. Huang, Q. Zhu, Farsighted risk mitigation of lateral movement using dynamic cognitive honeypots, in *International Conference on Decision and Game Theory for Security* (Springer, 2020), pp. 125–146
30. S. Mauw, M. Oostdijk, Foundations of attack trees, in *International Conference on Information Security and Cryptology* (Springer, 2005), pp. 186–198
31. S. Jha, O. Sheyner, J. Wing, Two formal analyses of attack graphs, in *Proceedings 15th IEEE Computer Security Foundations Workshop. CSFW-15* (2002), pp. 49–63
32. M. Rausand, A. Høyland, *System Reliability Theory: Models, Statistical Methods, and Applications*, vol. 396 (Wiley, 2003)
33. S. Contini, V. Matuzas, Analysis of large fault trees based on functional decomposition. Reliab. Eng. Syst. Saf. **96**(3), 383–390 (2011)
34. F. Baiardi, C. Telmon, D. Sgandurra, Hierarchical, model-based risk management of critical infrastructures. Reliab. Eng. Syst. Saf. **94**(9), 1403–1415 (2009)
35. D.W. Coit, E. Zio, The evolution of system reliability optimization, Reliab. Eng. Syst. Saf. **192**, 106259 (2018)
36. M. Todinov, Methods for analysis of complex reliability networks, in *Risk-Based Reliability Analysis and Generic Principles for Risk Reduction* (Elsevier, 2007), pp. 31–58
37. N. Leveson, *Engineering a Safer World: Systems Thinking Applied to Safety* (MIT Press, 2011)
38. D.W. Goodall, A new similarity index based on probability. Biometrics **22**(4), 882–907 (1966). [Online]. Available: http://www.jstor.org/stable/2528080
39. NIST SP 800–30: Guide for conducting risk assessments, National Institute of Standards and Technology, Gaithersburg, MD, Tech. Rep. (2012)
40. J. Fussell, E. Henry, N. Marshall, MOCUS: A computer program to obtain minimal sets from fault trees, Aerojet Nuclear Co., Idaho Falls, Idaho (USA), Tech. Rep. (1974)
41. W.S. Lee, D.L. Grosh, F.A. Tillman, C.H. Lie, Fault tree analysis, methods, and applications: A review. IEEE Trans. Reliab. **34**(3), 194–203 (1985)
42. A. Rauzy, Toward an efficient implementation of the MOCUS algorithm. IEEE Trans. Reliab. **52**(2), 175–180 (2003)

Chapter 3
Risk Mitigation Decisions

Abstract It is highly complex for organizations to navigate the emerging cyber-security landscape due to the lack of available decision-support tools. In this chapter, we present a systematic approach to supply chain risk mitigating decision-making in IoT systems and networks. The framework discovers relationships between suppliers and service providers across the different interconnected devices and provides an analysis of the associated cyber risks including the weakest and most vulnerable links. The decision-support engine allows for planning new system deployments from a supply chain viewpoint by recommending an optimized selection of suppliers.

3.1 Cost Effective Vendor Selection

The system model of I-SCRAM supports both risk analysis and mitigation decisions [1, 2]. Here, we discuss mitigation decisions as a series of related optimization problems that involve supply chain security. These problems in general are called *supplier choice problems* [3]. A supplier choice problem exists when, given a component security graph and universal supplier network as described above, an acquirer or integrator must choose a supplier from among several options while seeking to minimize risk within a certain budget. An optimal supplier choice must take into account the topologies of both networks: the supplier network and the component security graph. As IoT systems involve extensive complexity in both networks, risk optimization problems will be challenging to approach [4, 5]. Here we focus on providing a functional approach to risk optimization that can be applied both to self-contained systems and larger, system-of-systems problems.

Figure 3.1 shows a simple system with only two components. For the system to function, both components must function. Component c_1 has only one supplier option, a, whereas component c_2 may be provided by either supplier b or c. Stepping up to the top level of the diagram, two groups are represented, where A controls suppliers a and b and B controls supplier c. The supplier choice problem for this system has only two feasible solutions, represented as the left and right diagrams. Dotted black lines indicate a possible choice that is not taken, whereas solid red lines

T. Kieras et al., *IoT Supply Chain Security Risk Analysis and Mitigation*,
SpringerBriefs in Computer Science, https://doi.org/10.1007/978-3-031-08480-5_3

Fig. 3.1 Example scenarios of the supplier choice problem. Both represent feasible solutions where differing suppliers are chosen for component c_2, involving the system in different supplier groups

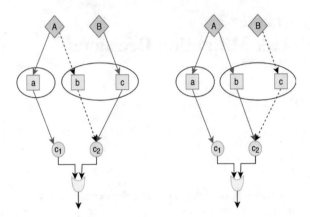

indicate the chosen supplier and which supplier network is involved. The optimal choice is found by choosing between suppliers b and c such that the resulting risk to the system is minimal, given a specified budget.

3.1.1   Strict Supplier Choice Problem

The first problem we approach directly aims to minimize system risk and serves as the basis for understanding the remaining supplier choice problems we discuss. This problem is formulated as a binary integer programming problem, with the decision variable $x = [x_{ij}]$, $i \in \{1 \ldots n\}$, $j \in \{1 \ldots m\}$, indicating which supplier should be chosen for each component. There are n components in the system, and m suppliers. The set X_i indicates the subset of suppliers that offer the component i. We consider various parameters for each supplier choice: the security risk of the component (r_{ij}), the cost of the component (c_{ij}) and the trust value of the supplier (t_j).

Definition 3.1 The *strict supplier choice problem* minimizes the general system risk function subject to the constraint of a specified budget. We formulate this nonlinear integer program as follows:

$$\min_{\mathbf{x}} \quad R(\mathbf{r}(\mathbf{x}, r, t)), \tag{3.1}$$

where

$$\mathbf{r}(\mathbf{x}, r, t) = \{r_i \mid r_i = \sum_{j=1}^{m} x_{ij} r_{ij} , i \in \{1, \ldots, n\}\}$$

$$\cup \; \{\bar{t}_j \mid \bar{t}_j = \sum_{i=1}^{n} x_{ij} \bar{t}_{ij} , j \in \{1, \ldots, m\}\}, \tag{3.2}$$

subject to

$$\sum_{i=1}^{n}\sum_{j=1}^{m} c_{ij}x_{ij} \leq b, \qquad b \in \mathbb{R}^{+}, \tag{3.3}$$

$$x_{ij} \in \{0, 1\}, \qquad i \in \{1, \ldots, n\}, j \in \{1, \ldots, m\}, \tag{3.4}$$

$$\sum_{j=1}^{m} x_{ij} \leq 1, \qquad i \in \{1, \ldots, n\}, \tag{3.5}$$

$$\sum_{x=1}^{n}\sum_{j \in X_i} x_{ij} \leq 1. \tag{3.6}$$

The objective function here directly computes the systemic risk as defined in Definition 2.3.1, given a vector of risk values for each entity in the system. In (3.1)–(3.2), the decision variable x is used to assemble a vector of risk values from component risks and supplier trusts. The budget constraint is expressed in (3.3), such that the sum of chosen costs not exceed the budget b. A single choice of supplier for each component is enforced by (3.5). The supplier choice must be among those offering the component, as required by (3.6).

For this optimization problem, a serious difficulty arises from the fact that the computation of the risk in the system requires an exhaustive discovery of the but the topography of the system graph depends on the decision. While the component security graph is static and its cutsets could be pre-computed, the choice of suppliers entails modifying the system graph. As in the simple example shown in Fig. 3.1, to choose supplier b entails that the system now has a single point of failure in controller A. Because supplier networks can be arbitrarily complex, a feasible approach to solving the supplier choice problem requires the use of heuristic measures that reduce the computational complexity of the problem. Our approach to this problem is to break the problem into two parts, as shown in Fig. 3.2. We now discuss each part in the following sections before combining them into a feasible approximation.

3.2 Supply Chain Diversification

Alongside satisfying budgetary constraints, it is also important to have diversification in the supply chain ecosystem where possible [6]. This prevents possibilities of potential collusion among potential attack vectors along with reducing the reliance of the system on a single or small number of vendors [4, 7]. We incorporate this feature into the decision framework by numerically quantifying the supplier involvement and then using it to make optimized selections [8].

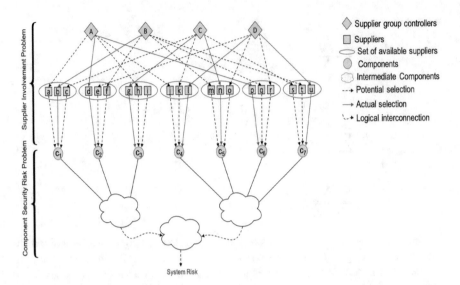

Fig. 3.2 An expanded supplier choice problem, divided into the two smaller problems of supplier involvement and component security risk

3.2.1 Component Security Risk Minimization Problem

The first step is to consider risk optimization in the component security graph.

> **Definition 3.2** We define the *Component security risk minimization problem* to be a supplier choice problem that minimizes component security risks using the risk importance measure from Sect. 2.3.1. The parameters and constraints are the same as the strict .
>
> $$\min_{\mathbf{x}} \ \sum_{j=1}^{m} \sum_{i=1}^{n} x_{ij} r_{ij} I_i. \tag{3.7}$$

Note that this problem does not consider suppliers, their trust value or the supplier topology. Consideration is given only to the risk value of the component choice on offer. The aim of any solution to assessing risk in a component security graph is to consider not only the risk values but also the topology and logical function of the corresponding component. We find the Birnbaum structural importance measure captures these features well, allowing for an approximation of system risk that is significantly simpler to compute.

Example 3.1 Because the use of the risk importance measure plays a significant role in our approach, it is helpful to explore the utility of the measure with an example. The example uses a system small enough that it is possible to compare the result of the approximation and the strict supplier choice problem. The system in question has only seven nodes, as presented in Fig. 3.3. Each component has seven possible suppliers. We use node 1 as the indicator node, with MOCUS [9] yielding the minimal cutsets [10] of this example system:

$$W = \{\{1\}, \{2\}, \{4, 5\}, \{3\}, \{6\}, \{7\}.$$

The above procedure for computing the Birnbaum structural importance for each node in the system yielded the results shown in Table 3.1. Both the strict supplier choice problem and the component security risk problem were solved for three sets of parameter cases described as follows.

1. **Case 1.** Linear risk:

$$c_{ij} = 80 - 10j \text{ for } i, j \in \{1, \ldots, 7\},$$
$$r_{ij} = 0.05j \text{ for } i, j \in \{1, \ldots, 7\}.$$

Fig. 3.3 A component security graph used to test the use of the risk importance measure. The output of Component 1 indicates a system failure event

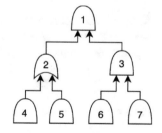

Table 3.1 The Birnbaum structural importance is calculated as the difference between system risk given two risk vectors

i	$R(s_i^1)$	$R(s_i^0)$	$I_i = R(s_i^1) - R(s_i^0)$
1	1.00000	0.95312	0.04688
2	1.00000	0.95312	0.04688
3	1.00000	0.95312	0.04688
4	0.98438	0.96875	0.01562
5	0.98438	0.96875	0.01562
6	1.00000	0.95312	0.04688
7	1.00000	0.95312	0.04688

2. **Case 2.** Logarithmic risk:

$$c_{ij} = 80 - 10j \text{ for } i, j \in \{1, \ldots, 7\},$$

$$r_{ij} = \frac{1}{\log c_{ij}} - 0.20 \text{ for } i, j \in \{1, \ldots, 7\}.$$

3. **Case 3.** Random linear risk

$$c_{ij} = 80 - 10j \text{ for } i, j \in \{1, \ldots, 7\},$$

$$r_{ij} = \text{Selected randomly from the interval}$$

$$[0.05j, 0.05j + 0.05] \text{ for } i, j \in \{1, \ldots, 7\}.$$

Both of the above two optimization problems were solved across the range of feasible budgets. After each instance was solved, yielding a decision matrix x, the general system risk function was computed based on vector of risk values indicated by x. This process was repeated for the three sets of parameters. The resulting risks are shown in Fig. 3.4. We find that the use of the risk importance measure functions well in minimizing system risk without the direct calculation of the system risk function.

3.2.2 Supplier Involvement Minimization Problem

Although the component risk minimization problem performs well, it does not consider the topology of the supplier network. The following problem is intended to capture this second portion of the strict supplier choice problem.

Fig. 3.4 Risk importance evaluation results. By minimizing the sum of risk importance values, the overall system risk is approximately minimized

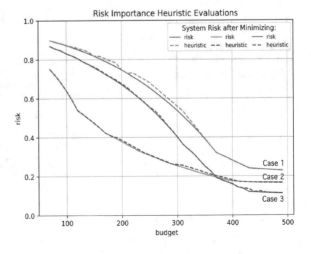

Definition 3.3 We define the *supplier involvement minimization problem* as a supplier choice problem that considers only supplier involvement as defined in Sect. 2.3.2. Using the same parameters and constraints as the strict supplier choice problem, this problem is formulated as follows:

$$\min_{\mathbf{x}} \sum_{k=1}^{K} \left(\sum_{j \in G_k} \sum_{i=1}^{n} x_{ij} I_i \right)^2 \bar{t}_k. \tag{3.8}$$

Example 3.2 Figure 3.1 illustrates a system suitable as an example of the supplier involvement minimization problem.

In this example there are two supplier groups: $G_1 : \{a, b\}$, $G_2 : \{c\}$. Let the trust values of the groups be: $t_A : 0.9$, $t_B : 0.9$. The two feasible solutions are represented by the decision vectors: $x_1 = (1, 0, 1)$, $x_2 = (1, 1, 0)$. Using the above formula for supplier involvement yields the following values: $SI_1 = 0.2$, $SI_2 = 0.4$. The minimal cutsets for each feasible solution are:

$$W_1 = \{\{c_1, c_2\}, \{a, c\}, \{a, c_2\},$$
$$\{c, c_1\}\{A, B\}, \{A, c\}, \{A, c_2\}, \{B, a\}, \{B, c_1\}\}$$
$$W_2 = \{\{c_1, c_2\}, \{a, b\}, \{a, c_2\}, \{b, c_1\}, \{A\}\}$$

Using the system risk function from Sect. 2.3.1 and taking all component risk values as zero yields the following results:

$$R_1 = 0.01$$
$$R_2 = 0.1$$

In this small example, we see that minimizing the supplier involvement measure yields a result that corresponds to the minimal system risk. We note this example to illustrate the use of the supplier involvement measure, while further noting that the measure's inherent trade-off between supplier grouping and supplier trust should not be expected to minimize risk in every case. Consequently, there is need to adjust the weighting of various sources of risk based on an organization's priorities.

3.2.3 Relaxed Supplier Choice Problem

Our approach to reducing the complexity of the supplier choice problem has been to split the problem into two parts and give each an appropriate heuristic measure. At this point we can combine the two in order to more feasibly approach the supplier choice problem.

Definition 3.4 We define the *relaxed supplier choice problem* as an approximation of the strict supplier choice problem employing both the supplier involvement measure and the risk importance measure. The constant α is used to weight the supplier involvement measure, and the parameters and constraints are the same as the strict supplier choice problem. The objective function is as follows:

$$\min_{\mathbf{x}} \sum_{j=1}^{m} \sum_{i=1}^{n} x_{ij} \hat{r}_{ij} I_i + \alpha \sum_{k=1}^{K} \left(\sum_{j \in G_k} \sum_{i=1}^{n} x_{ij} I_i \right)^2 \bar{t}_k. \tag{3.9}$$

The first term characterizes component security risk, based on the risk importance measure defined in Definition 2.3.1. The second term, weighted by α, adds a penalty based on supplier involvement as defined in Sect. 2.3.2. Recall that $\hat{r}$ stands for the adjusted component security risk, i.e., the security risk taking into account the trust value of the associated supplier. Because the relaxed supplier choice problem approximates the strict supplier choice problem, it must consider not only component risk but also supplier trust.

With larger values of α, the minimized solution avoids choices that consist of groups of suppliers across important components. However, this comes at the cost of potentially choosing an isolated supplier that brings a higher risk value. As is shown in the case study below, care must be taken when choosing α. While we consider that a modest value generally contributes to the minimization of general risk, we note that the object of our study is not simply the minimization of general risk but the construction of systems that are not beholden to significant involvement by groups of suppliers. If this goal is less significant to an organization, a minimal or zero value for α is appropriate.

3.3 Case Study and Results

To illustrate the use of I-SCRAM for risk analysis and mitigation decisions, we consider a case study of an autonomous vehicle system. The case study is based on a simplified system model and assumes that the vehicle has been designed already to meet functional requirements. In the sequel, we first describe the simulation setup and then provide the results based on selected scenarios.

3.3.1 Simulation Setup

The simulation setup has two main parts as shown in the block diagram in Fig. 3.2. The first part relates to component interconnections while the second part relates to the configuration of the supply chain. The principal security risk we model involves attacks against the availability of safety critical components. The system graph in Fig. 3.5 represents a hypothetical result of such a security analysis. It involves 15 different components such as 'steer_act', 'accel_act', 'brake_act', etc., and their logical connections. Each minimal cut on the graph represents a complete attack. We assume here that a complete attack compromises the availability of any of the three terminal actuators.

The supplier choice problem for this system involves choosing a supplier for each of the 15 components. For this case study we have chosen the following scenario for the supplier network: Each component may be acquired from one of three suppliers; No supplier offers more than one component; Of the 45 suppliers, there are five groups of five suppliers, with each group having a controller; The remaining 20 suppliers do not have a group controller. An illustration of the supplier hierarchy is shown in Fig. 3.6. Although the supplier network topology described above is kept unmodified, parameters for risk, cost, supplier trust and group controller trust are varied across several different cases. In all cases, riskier components cost less. Generally, the cases range from simpler to more complex, with details of each described below along with results and discussion.

A particular area of interest is the weighting of the two terms in the objective function in (3.9), controlled by the constant α. As α approaches zero, the supplier involvement measure contributes less and the resulting risk is more sensitive to group controller trust. When running experiments on varying budgets, each instance

Fig. 3.5 Component security graph for an autonomous vehicle. System security is indicated by the top node 'Indicator'. A security failure of any of the actuators causes an incident. Suppliers are not shown here

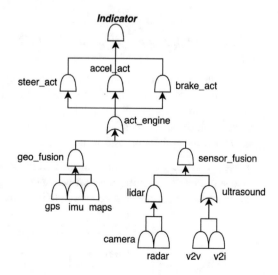

Fig. 3.6 Supply chain
hierarchy used for
simulations. The 15
components are represented
by nodes on the right while
the possible suppliers are
represented by nodes in the
middle and group controllers
are represented by nodes on
the left

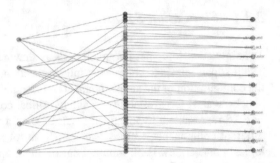

was solved with values of $\alpha \in \{0, 0.001, 0.01, 0.1\}$, though we present here only a
subset of the results that illustrate important aspects of the study.

The direct calculation of the system risk function entails a discovery of the
minimal cutsets after the composition of a component security graph with the
chosen supplier network. Because this is difficult for problems of significant size,
we have developed a Monte Carlo simulation framework that provides an empirical
approximation of the general security risk function. This simulation provides a
method to evaluate the performance of the relaxed supplier choice problem across
different budgets and values of α. In each iteration of the simulation, a supplier is
either normal or malicious based on a Bernoulli trial with its trust value. Similarly,
a component is either functional or non-functional based on a Bernoulli trial with
its risk value. The simulation ($n = 10^4$ iterations) results in two measures of system
risk: the probability of a general security failure (*general risk*), and the probability
of a critical security failure that is caused only by group controllers (*group risk*). We
expect that higher values of α should yield systems with generally decreasing group
risk, while lower α values yield generally decreasing general risk.

3.3.2 Example Scenarios and Results

We consider several different scenarios for components, their cost structures, as
well as trust level of suppliers for evaluating our framework. In each scenario, the
parameters used are provided explicitly. The supplier choice problem in each case
is solved using mixed integer non-linear programming . The resulting risk is then
plotted for varying budgets and values of α.

1. **Case 1** tests the scenario of minimal complexity. All risk values are zero except
   one supplier group. It implies that we have a single malicious supply chain actor
   in the system that needs to be avoided. The parameters selected are as follows:

$$r_{ij} = 0.0 \text{ for } i \in \{1, \ldots, n\}, \ j \in \{1, \ldots, m\},$$
$$t_j = 1.0 \text{ for } j \in \{1, \ldots, m\},$$

$$t_k = 1.0 \text{ for } k \in \{1, \ldots, K\},$$

$$t_k = 0.8 \text{ for } k = 1.$$

In this case, we expect the first term of the objective function always to be zero, with there being only one group that contributes any risk. Results are shown in Fig. 3.7a, with $\alpha = 0.1$. With a budget $b > 1675$, the objective function can reach zero. The simulation shows that this yields a system with a risk of zero. Because the only source of risk in this case is from a group, the general and group risks are identical.

2. **Case 2** features two risky groups, but maintains no other source of risk. The parameters are:

$$r_{ij} = 0.0 \text{ for } i \in \{1, \ldots, n\}, j \in \{1, \ldots, m\},$$

$$t_j = 1.0 \text{ for } j \in \{1, \ldots, m\},$$

$$t_k = 1.0 \text{ for } k \in \{2, \ldots, K\},$$

$$t_k = 0.8 \text{ for } k \in \{1, 2\}.$$

Results are shown in Fig. 3.7b. As in case 1, here simulated general risk and group risk are identical, and as the objective function reaches zero, so also do the risks. Since there are two risky groups and therefore ten supplier-component choices that entail some risk, optimization across increasing budgets results in a successively decreasing objective function that eventually eliminates the risk as the budget gets large enough.

3. **Case 3** continues increasing the complexity of the scenario. All group controllers have a high, randomly generated trust value $t < 1.0$. No other sources of risk are yet considered. The parameters are:

$$r_{ij} = 0.0 \text{ for } i \in \{1, \ldots, n\}, j \in \{1, \ldots, m\},$$

$$t_j = 1.0 \text{ for } j \in \{1, \ldots, m\},$$

$$t_k = \mathcal{N}(\mu = 0.98, \sigma = 0.01) \text{ for } k \in \{1, \ldots, K\}.$$

In this case, we notice that the objective successively decreases as a higher budget is available. However, it does not go to zero as in the previous cases. This is because of multiple sources of risk present in the supply chain (Fig. 3.8).

4. In **Case 4**, both suppliers and groups introduce risk, but components themselves still do not. The parameters are as follows:

$$r_{ij} = 0.0 \text{ for } i \in \{1, \ldots, n\}, j \in \{1, \ldots, m\},$$

$$t_j = \mathcal{N}(\mu = 0.98, \sigma = 0.01) \text{ for } j \in \{1, \ldots, m\},$$

$$t_k = \mathcal{N}(\mu = 0.98, \sigma = 0.01) \text{ for } k \in \{1, \ldots, K\}.$$

Fig. 3.7 Results for simple
cases. General and group
risks overlap and decrease
with higher budgets

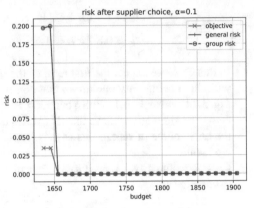

((a)) Case 1 Results, $\alpha = 0.1$

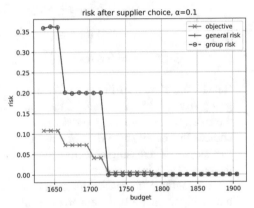

((b)) Case 2 Results, $\alpha = 0.1$

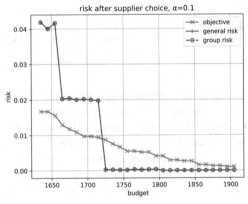

((c)) Case 3 Results, $\alpha = 0.1$

Fig. 3.8 Case 4 results begin
to show the divergence
between group and general
risk. Weighting within the
objective function assigns the
priority. Both risks decrease
but the non-prioritized risk
shows instability

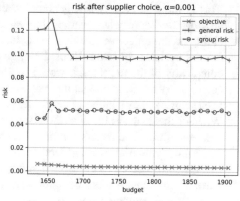

((a)) Case 4 Results, $\alpha = 0.001$

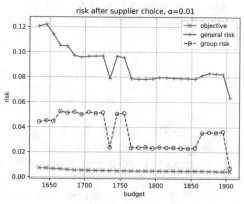

((b)) Case 4 Results, $\alpha = 0.01$

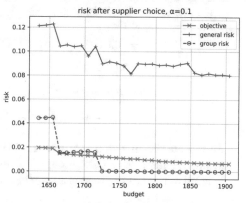

((c)) Case 4 Results, $\alpha = 0.1$

This is the first case where general system risk and group risk diverge. It is also the first case where setting the value of α is non-trivial and affects the ability of the objective function to capture a consistently minimized risk. This is not surprising for two reasons: First, given the use of heuristics to reduce the complexity of the problem, it would be unusual to find risk to strictly decrease across every arbitrary budget interval. Second, if supplier group involvement has been prioritized by a high α, then riskier components or suppliers may be chosen. Whether this is desirable is at the discretion of the organization conducting the risk modelling. The results for case 4 show that the value of α affects whether general or group risk is prioritized as the budget increases. The non-prioritized risk generally decreases but with significant instability.

5. **Case 5** represents a more comprehensive, typical scenario where there is risk from groups, suppliers and components, but all risks are small and similar. Component risk is set by a linear function of its cost implying that the component risk decreases linearly as its cost increases. There are no significantly risky entities. The parameters are specified as follows:

$$r_{ij} = f(c_{ij}) \text{ for } i \in \{1, \ldots, n\}, j \in \{1, \ldots, m\},$$

$$t_j = N(\mu = 0.98, \sigma = 0.01) \text{ for } j \in \{1, \ldots, m\},$$

$$t_k = N(\mu = 0.98, \sigma = 0.01) \text{ for } k \in \{1, \ldots, K\}.$$

Results in Fig. 3.9b show that minimizing the objective function tends to minimize system risk. However, with an α of 0.1, the resulting system is optimized to avoid group risk and so general risk occasionally increases. By contrast, Fig. 3.9a shows the results of optimizing for minimal general risk, which decreases in a more stable manner at the expense of a generally higher level of group risk.

6. **Case 6** represents a scenario where the supplier network contains some unusually risky groups. These risky groups have controllers that are trusted with comparatively low values, i.e., 80% of the other groups. The parameters are:

$$r_{ij} = (0.01, 0.04) \text{ for } i \in \{1, \ldots, n\}, j \in \{1, \ldots, m\},$$

$$t_j = N(\mu = 0.98, \sigma = 0.01) \text{ for } j \in \{1, \ldots, m\}$$

$$t_k = N(\mu = 0.98, \sigma = 0.01) \times 0.8 \text{ for } k \in \{1, 2\},$$

$$t_k = N(\mu = 0.98, \sigma = 0.01) \text{ for } k \in \{3, \ldots, K\}.$$

Results are shown with $\alpha = 0.1$ in Fig. 3.9c. The lowest budgets are constrained to involve these very risky groups, but in both cases the simulated risks are reduced as each risky group is able to be eliminated from the system.

Fig. 3.9 More complex cases
confirm the risk minimizing
performance along with the
effect of prioritizing general
and group risk

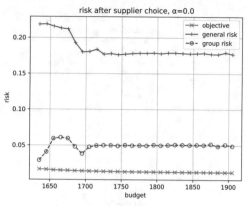

((a)) Case 5 Results, $\alpha = 0.0$

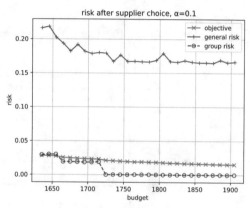

((b)) Case 5 Results, $\alpha = 0.1$

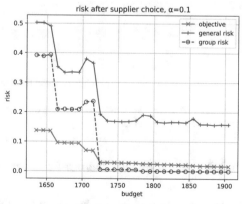

((c)) Case 6 Results, $\alpha = 0.1$

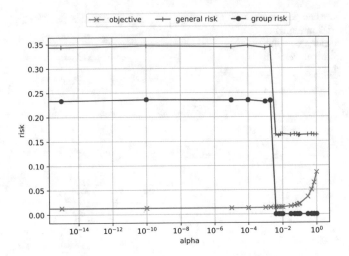

((a)) Case 6 Alpha Experiment, $b = 1850$

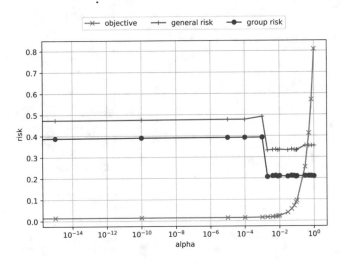

((b)) Case 6 Alpha Experiment, $b = 1675$

Fig. 3.10 With constant budget, α is varied to show the effect of weighting and the presence of state changes in the problem

3.3.3 Supplier Involvement Experiments

Throughout these cases, we have seen that the value of α occasionally affects the results in significant ways. To explore this further, we present two experiments where the budget is kept constant and only α varies. The other parameters are taken from Case 6.

Figure 3.10a shows the objective function and the two simulated risk results for case 6 at the budget of 1850, with a variety of choices of $\alpha \in [0, 1]$. This high budget should require relatively few difficult choices. The results show that in this case a clear transition occurs when $\alpha = 0.004$, after which the assessment of supplier group involvement dominates the objective function and drives down risk significantly. The reduction of group risk to zero after this transition further indicates that this type of risk accounts for the significant decrease in overall risk. Lastly, we notice that increased values of $\alpha > 0.004$ do not further decrease risk. These larger values increase the value of the objective function; however, because the group risk has already been driven to zero, no further gains are achieved by further weighting it.

The second α experiment was conducted with a lower budget of $b = 1675$, chosen to be near a potential state transition as found in the results for case 6. As shown in Fig. 3.10b, the effect of weighting can be detected by a small decrease in general risk. The higher values of alpha represent prioritization of group risk, which is successfully decreased. Yet, we note the expected increase in general risk that results from this choice of priorities.

3.4 Conclusion

In this chapter, we have presented a model for understanding the role of suppliers in IoT system security risk assessments and argued for the need to take a component centered approach to risk assessment when considering supply chain threats. With the developed model we have presented measures for risk analysis and approached optimization of budget-constrained supplier choices. Because IoT systems have complex topologies and may be involved with a large number of suppliers, we have developed an approximation of the supplier choice problem that can be more easily scaled to approach the complex, system-of-systems scenarios of IoT systems. Likewise, our approximation is capable of prioritizing various sources of risk according to the needs of the organization. In a case study, we have shown that our approach successfully decreases risk as the budget constraint increases.

References

1. C.K. Wu, K.F. Tsang, Y. Liu, H. Zhu, Y. Wei, H. Wang, T.T. Yu, Supply chain of things: A connected solution to enhance supply chain productivity. IEEE Commun. Mag. **57**(8), 78–83 (2019)
2. T. Kieras, J. Farooq, Q. Zhu, I-SCRAM: A framework for IoT supply chain risk analysis and mitigation decisions. IEEE Access **9**, 29827–29840 (2021)
3. H.P. Ho, The supplier selection problem of a manufacturing company using the weighted multi-choice goal programming and MINMAX multi-choice goal programming. Appl. Math. Model. **75**, 819–836 (2019). [Online]. Available: https://www.sciencedirect.com/science/article/pii/S0307904X19303610

4. *Network Defense Mechanisms Against Malware Infiltration* (Wiley, 2021), ch. 8, pp. 97–124 [Online]. Available: https://onlinelibrary.wiley.com/doi/abs/10.1002/9781119716112.ch8

5. J. Farooq, Q. Zhu, *Internet of Things-Enabled Systems and Infrastructure* (Wiley, 2021), ch. 1, pp. 1–8. [Online]. Available: https://onlinelibrary.wiley.com/doi/abs/10.1002/9781119716112.ch1

6. H.Y. Mak, Z.J. Shen, Risk diversification and risk pooling in supply chain design. IIE Trans. **44**(8), 603–621 (2012)

7. M.J. Farooq, Q. Zhu, Modeling, analysis, and mitigation of dynamic botnet formation in wireless IoT networks. IEEE Trans. Inf. Forens. Secur. **14**(9), 2412–2426 (2019)

8. J. Farooq, Q. Zhu, *Resource Management in IoT-Enabled Interdependent Infrastructure* (Wiley, 2021), ch. 2, pp. 9–13. [Online]. Available: https://onlinelibrary.wiley.com/doi/abs/10.1002/9781119716112.ch2

9. A. Rauzy, Toward an efficient implementation of the MOCUS algorithm, IEEE Trans. Reliab. **52**(2), 175–180 (2003)

10. M. Todinov, Methods for analysis of complex reliability networks, in *Risk-Based Reliability Analysis and Generic Principles for Risk Reduction* (Elsevier, 2007), pp. 31–58

Chapter 4
Policy Management

Abstract Supply chain security has become a growing concern in the security risk analysis of IoT systems. Their highly connected structures have significantly enlarged the attack surface, making it difficult to track the source of the risk posed by malicious or compromised suppliers. This chapter presents a system-scientific framework to study the accountability in IoT supply chains and provides a holistic risk analysis technologically and socio-economically. We develop stylized models and quantitative approaches to evaluate the accountability of the suppliers. Two case studies are used to illustrate accountability measures for scenarios with single and multiple agents. Finally, we present the contract design and cyber insurance as economic solutions to mitigate supply chain risks. They are incentive-compatible mechanisms that encourage truth-telling of the supplier and facilitate reliable accountability investigation for the buyer.

4.1 Introduction

Supply chains play a critical role in the security and resilience of IoT systems and affect many users, including small- and medium-sized businesses and government agencies. An attacker can exploit vulnerabilities of a vendor in the supply chain to compromise the IoT system at the end-user. The recent SolarWinds attack is an example of an attack that has resulted in a series of data breaches at government agencies. One seller of the Microsoft Cloud services was compromised by the attacker, allowing the attacker to access the customer data of its re-sellers. Once the attacker established a foothold in SolarWind's software publishing infrastructure after getting access to SolarWind's Microsoft Office 365 account, he stealthily planted malware into software updates that were sent to the users, which include customers at US intelligence services, executive branch, and military.

The infamous Target data breach in 2013 is another example of supply-chain attacks. The attacker first broke into Target's main data network through ill-protected HVAC systems. The attacker exploited the vulnerabilities in the monitoring software of the HVAC systems, which shared the same network with the data services. It led to a claimed total loss of $290 million to data breach-related

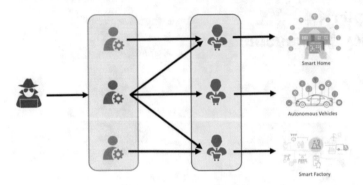

Fig. 4.1 Supply-chain attacks: An attacker first attacks a vendor, who sells the users compromised products. They act as Trojans inside the user's system and stealthily manipulate it

fees [1, 2]. The supply-chain attacks would become increasingly pervasive in IoT systems. Consider a next-generation industrial manufacturing plant equipped with IoT devices that are supported by third-party vendors . The software and the hardware of these devices can be trojanized. As a result, the attacker disrupts the manufacturing plant, which can create a shortage of essential products (e.g., pharmaceutical products, COVID19 vaccines, and gasoline) and lead to grave repercussions in the nation's supply chain. Illustrated in Fig. 4.1, the supply chain attacks can go through multiple stages of the supply chain from the source of the attack to the targeted users or systems.

Risk-based approaches presented in Chaps. 2 and 3 have been used to guide the procurement and design decision-making process [3–7]. This kind of approach offers risk measurement, rating tools, and compliance checking to identify and rank the vendors by their risk criticality. It is a useful preventive measure that provides a transparent understanding of the security posture in the products, systems, and services of the end-users and helps mitigate the risks prior to the procurement contracts and continuous product development. Cyber resilience complements this measure. It shifts the focus from prevention to recovery by creating a cyber-resilient mechanism to reconfigure the IoT system adaptively to the uncertainties of adversaries and maintain critical functions in the event of successful attacks.

Many private sectors have for years prioritized efficiency and low cost over security and resilience. In addition, they are agnostic to where these technologies are manufactured and where the associated supply chains and inputs originate. This common practice has resulted in enlarged attack surfaces and many unknown and unidentified threats in the IoT systems. A healthy ecosystem of vendors and suppliers is pivotal to secure and resilient IoT systems. One challenge is that the IoT supply chain is becoming globalized. Manufacturers and material suppliers are geographically diverse, thus increasing the uncertainties and the vulnerabilities of the end-user IoT systems. It is critical to check the compliance of the products from the global supply chain to determine whether they would increase the cyber risk of the IoT users.

One way to improve the health of the IoT supply chain is to design an IoT system with built-in security and resilience mechanisms. For example, the integration of cyber deception [8–10] into IoT systems provides a proactive way to detect and respond to advanced and persistent threats. Game-theoretic methods [11–13] and reinforcement learning techniques [14–16] have been used to provide a clean-slate approach to designing cyber resilient mechanisms in response to supply-chain attacks.

Apart from the technological solutions, accountability and insurance are the socio-economic ones that can be used to improve the cyber resilience of IoT end-users. Accountability, in general, is the ability to hold an entity, such as a person or organization, responsible for its actions. An accountable system can identify and punish the party or the system component that violates the policy or the contract. By creating accountable IoT supply chains, we create an ecosystem where each supplier invests in cybersecurity to reduce the cyber risks at each stage of the supply chain. A supplier would be held accountable if the failures of the end-user system are attributed to it. Accountability establishes a set of credible incentives for the suppliers and elicits desirable behaviors that mitigate the cyber risks. Accountability can be viewed as part of the cyber resilience solutions succeeding the technological solutions, especially when the technological resilience measures do not prevent the damages.

Insurance is another risk management tool [17] to protect the end-users from cyber attacks and failures by transferring their residual risk from an entity to a third party through an insurance contract. It is the last resort when an IoT system cannot be perfectly accountable; i.e., there is inadequate evidence to hold any one of the suppliers accountable, or when the defects in the user's design lead to unanticipated consequences. The residual risks would be evaluated by an underwriter and the coverage can include the losses that arise from ransomware and data theft or incidents caused by failures of IoT devices. Figure 4.2 shows the relationships between preventive cyber measures and resilient cyber measures. The cyber-resilient mechanisms include the technological real-time resilience measures as well as accountability and insurance solutions. They constitute a holistic socio-technical solution to protect the IoT systems from supply-chain threats.

Both accountability and insurance provide an additional layer of protection that reduces the risks of IoT users. Accountability and insurance are system-level issues. We need to take a system-scientific and holistic approach to understand their role in IoT systems and supply chains, which would lead to an integrative socio-technical solution for supply chain security. This chapter provides a quantitative definition to measure and assess the accountability in the IoT supply chain that pertains to the system design, procurement contracts, as well as, vendor description. Despite the focus of the chapter on cybersecurity issues, the definition of accountability can be extended and used for general contexts of supply chain disruptions caused by natural disasters and the defects in the products.

Game theory naturally provides a framework that captures the incentives and penalties through utility functions for multiple interacting agents. It has been widely used in cybersecurity for the modeling between an attacker and defender in many

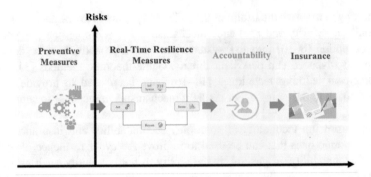

Fig. 4.2 The IoT supply chain can be protected using preventive measures which include compliance checking and auditing. The supply chain resilience can be enhanced by building real-time resilience measures (e.g., detection, adaptation, and reconfigurations). The residual risk as a result of the preventive and real-time resilience measures can be further mitigated by accountability and insurance mechanisms. Accountability is designed to attribute the violations to the suppliers, who will be penalized based on the contract. Insurance is another mechanism to transfer the remaining risks to a third party through an insurance contract. The multi-tier solutions from preventive measures to insurance are interdependent and they create consolidated protection of our IoT supply chain ecosystem

scenarios, including intrusion detection systems [18–21], wireless communications [22–27], and cyber deception [8, 12, 28–31]. It has also been used to harden the security and resiliency of cyberphysical systems, including critical infrastructures [32–36], industrial control systems [11, 37–41], and IoTs [42–46].

One important branch of game theory is the mechanism design theory [47, 48] , which explicitly provides a quantitative approach to create a reward and penalty mechanism to elicit desirable behaviors at equilibrium. The violations from the desired behaviors would be disincentivized or punished, while the compliance with the rules would be incentivized or rewarded. In this chapter, we leverage these features of game theory to create computational accountability and insurance framework for IoT systems and their supply chain.

Accountability is a system-level issue that encompasses detection and attribution of the violations or anomalies, multi-agent interactions, asymmetric information, and feedback. Game-theoretic methods provide a baseline for a system-scientific view for accountability. We build a system scientific framework that bridges game theory, feedback system theory, detection theory, and network science to provide a holistic view toward accountability in IoT supply chains. The framework proposed here can be applied to understand accountability in general.

One extension of this chapter is to investigate the concept of collective accountability, where multiple agents are held accountable for the violations. One advantage of such accountability mechanisms is the convenience in identifying the entities to be held accountable and the implementation of the penalties. The disadvantage is that they are not targeted and entities that are not directly linked to the violation of the failures would be also punished.

4.2 Literature Review

Accountability has been studied in many different contexts in computer science [49–51]. Künnemann et al. in [52] have studied accountability in security protocols. Accountability is defined as the ability of a protocol to point to any party that causes failure with respect to a security property. Zou et al. in [53] have proposed a service contract model that formalizes the obligations of service participants in a legal contract using machine-interpretable languages. The formalism enables the checking of obligation fulfillment for each party during service delivery and holds the violating parties for the non-performance of the obligations. The definition of accountability in these works aligns with the definition in this chapter. An accountable system has the ability to check and verify compliance with the requirements in the agreement and identify the non-conforming behaviors and their parties.

There are several game-theoretic models that are closely related to accountability. For example, inspection games [54–56] are one class of games where the inspector determines a strategy to examine a set of sampled items from a producer to check whether the producers of the goods violated the standards. The producer aims to set a production strategy to minimize the detection probability while minimizing the cost of maintaining high standards. The inspection games have been used in many contexts such as patrolling, cybersecurity, and auditing. Blocki et al. in [57] have studied a class of audit games in which the defender first chooses a distribution over n targets to audit and the attacker then chooses one of the n targets to attack. It is better for the defender to audit the attacked target than an unattacked target, and it is better for the attacker to attack an unaudited target than an audited one. Rass et al. in [58, 59] have studied a multi-stage cyber inspection game between a network system defender and an advanced persistent threat (APT) attacker. The defender needs to choose an inspection strategy to detect anomalies at different layers of the networks. The attacker's goal is to stay stealthy and find strategies to evade the detection and compromise the target.

Utility-theoretic approaches are useful to capture the incentives of the participants in an agreement and their punishment. In [51], Feigenbaum et al. have formalized the notion of punishment using a utility-theoretic, trace-based view of system executions. Violation is determined based on the traces of the participants. When there is a violation, the participant is punished. This punishment is captured through a decrease in the utility, relative to the one without the violation. This approach to punishment is often seen in the literature of mechanism design [48, 60]. The designer first announces a resource allocation rule and a payment or punishment rule. The participants in the mechanism know the rules and determine the messages that they send to the designer. An incentive-compatible mechanism is one in which the participants will truthfully reveal their private information through the message under the allocation and the punishment rules. In other words, no participants have incentives to lie about their private information under an incentive-compatible

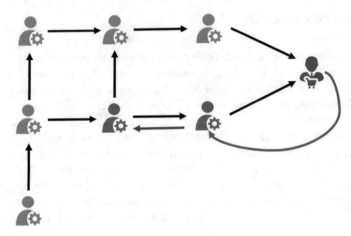

Fig. 4.3 Supply chain accountability: the buyer of the product can identify the supplier of a component who violates the policies or the contracts. The buyer can then use the contract to penalize the identified supplier. The supplier can attribute the violation to his supplier. It is called multi-stage accountability

mechanism. Mechanism designs have been used in many disciplines to study pricing of resources [61–63], create security protocols [45, 64], recommend policies [10], and design services [17, 65]. The framework that we present in this chapter is built on the mechanism design approach. The utility-theoretic approach conveniently captures the incentives of the suppliers and their behaviors. Furthermore, the mechanism-design approach naturally creates a punishment mechanism to create incentives for truthful behaviors. This type of behavior can be generalized to compliant behaviors in supply chain agreements and contracts.

Our framework builds on this approach and bridges the accountability gap by incorporating the detection mechanism that enables the designer to detect and attribute the non-compliant behaviors. In addition, our framework distinguishes from prior works in accountability by focusing on accountability in system engineering. This problem is instrumental in the development of large-scale IoT systems, where the building blocks of the IoT systems are manufactured or designed by third parties. We integrate the critical component of engineering designs into the accountability problem for IoT systems. The system designs can contribute to accountability. A design is called *transparent* if it helps identify the cause of the accidents; otherwise, a design makes the accountability inconspicuous. Figure 4.3 illustrates the concept of accountability. A user can use his observed information to identify the immediate cause of accidents or malfunctions. The seller who has been identified as the cause can further identify the further cause of the event. In this way, the source of the attack can be sequentially identified stage-by-stage through a chain of accountability efforts.

4.3   Accountability Models in IoT Supply Chain

4.3.1   Running Examples

We introduce two running examples which will be used in later discussions for illustrations.

Example I: Uber Autonomous Vehicles The Uber incident in Tempe, Arizona is another example of accountability of autonomous vehicles. A pedestrian was struck by an Uber self-driving vehicle with a human safety backup driver in the driving seat. The fatality is caused by the failure of the software system which fails to recognize the pedestrian. Sensor technologies, including radar and LiDAR, are sophisticated enough to recognize objects in the dark. Evidence has shown that the pedestrian was detected 1.3 s before the incident and the system determined that emergency braking was required but the emergency braking maneuvers were not enabled when the vehicle is under computer control. The design of the software system is accountable for the death of the pedestrian.

Example II: Ransomware Attack on Smart Homes A smart home consists of many modern IoT devices, including lighting systems, surveillance cameras, autonomous appliance control systems, and home security systems. The components of each system are supplied by different entities. Smart home technology integrates the components and creates a functioning system that will sense the home environment, make online decisions, and control the system. The camera is accountable if the home security system does not respond to the burglary adequately due to a camera failure. There is an increasing concern about ransomware attacks. Accountability enables the homeowner to mitigate the impact of the ransomware by attributing the attack to a supplier of the IoT devices.

Illustrated in the two examples, IoT supply chain security has a significant impact on the private sector and its customers. Several technologies have been proposed to track the integrity of the supply chain to provide real-time monitoring and alerts of tampering and disruptions. They provide a tool to monitor, trace, and audit the activities of all participants in the supply chain and ensure that the contractually defined Service Level Agreements (SLAs) are followed. The essence of the technologies is to create transparency and situational awareness for the companies. However, the software and hardware tampering is much harder to monitor and track than the physical one. As a result, it creates information asymmetry where the buyers or the systems do not have complete information about their suppliers. As in the Target and the SolarWinds attacks, an attacker can get access to the system through a compromised third-party vendor. It would require proactive security mechanisms to detect and respond to the exploited vulnerabilities. We have seen the emerging applications of cyber deception [8, 9, 66] and moving target defense [67–69] in both software and hardware to reduce the information

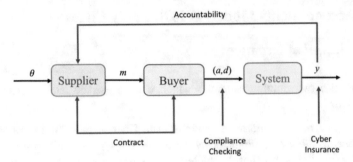

Fig. 4.4 A supplier of type θ provides a description m of the product to a buyer who will make a procurement decision a. The system designer develops a design d to integrate all the components to form a functioning system. The system, as a result, yields an observable performance y. The supply chain is said to be accountable if the malfunction of the system can be attributed to the supplier who has misled the system designer. The supply chain risk can be mitigated at three stages. The first stage is compliance checking before the procurement. The buyer can check whether the description of the product complies with the standards, regulations, and requirements. The second stage is the contracting stage. The buyer can make a contract that specifies the penalty or the consequences if the supplier does not fully disclose the product information. It will allow the buyer to hold the supplier accountable when the root cause of the malfunction is at the supplier. The third stage is cyber insurance. The buyer can purchase cyber insurance to mitigate the financial impact of the malfunction. The financial risk is partially transferred to the insurer

asymmetry and create proactive mechanisms for detection. They are tools that contribute to real-time resilience measures as illustrated in Fig. 4.4 and provide inputs for accountability in the next stage.

4.3.2 System Modeling

In this section, we provide a stylized model and a quantitative approach to accountability. Figure 4.4 describes three stages of interactions. At the first stage, a supplier interacts with a buyer to agree on an SLA contract. The supplier is characterized by the private information $\theta \in \Theta$, which is a true description of the product of the supplier. For example, the supplier is aware of the true security level and investment in the product but may not disclose the information to the buyer. The supplier sends the buyer a message $m \in M$, which is the informed description of the product. The description can prevaricate, hide, or sometimes lie about the security information that would be useful in the procurement decisions. We say that the supplier truthfully reports the product when $\theta = m$; otherwise, we say that the supplier misinforms the buyer. This misinformation can be unintended or intentional. In the case of intentional behaviors, the supplier sends a manipulative message when he knows his true type. For example, some foreign suppliers do not fully disclose the information of their product with the aim to attract US customers due to its low cost. In the case of unintended behaviors, the supplier may not be aware of the vulnerabilities of the product and sends a description based on his perceived information. In this case, we can assume that the private information θ is a function $\rho : \Theta \times \mathcal{W} \mapsto \Theta$ of the truth and uncertainties, i.e., $\theta = \rho(\theta_t, w_t)$, where $\theta_t \in \Theta$ is the true value unobservable by the supplier and $w_t \in \mathcal{W}$ is the bias, modeled as a random variable, unknown to the supplier. This bias can be interpreted as the uncertainties introduced by nature or a stealthy attacker that has unknowingly changed the security attributes of the product. In both cases of unintended and intentional behaviors, it is sufficient to assume that the type known to the supplier is θ.

Based on the product description m, the buyer can make purchase decisions. Let $a = 1$ denote the decision of adopting the product of the vendor and $a = 0$ otherwise. The decision rule $\alpha : \mathcal{M} \mapsto [0, 1]$ yields the probability of purchase based on the received description, i.e. $\alpha(m) = \Pr(a = 1|m)$. This can be interpreted as the purchase preference from historical records. If the buyer decides to adopt the product, then he determines how the product is designed and integrated into the system. Here, we assume that the user and the designer belong to the same organization and hence the procurement and design decisions are made jointly. In other words, the user and the designer can be viewed as the same decision entity who coordinates the design and procurement. In practice, the engineers design the systems and send the procurement department the specifications and requirements for the needed materials and components.

An IoT system consists of many components. We can classify the components into five major categories: sensing, computation, control, communications, and hardware. The sensing component allows the system to provide information about the environment, for example, the LiDAR and temperature sensors. The computation units provide functions and services for information processing and computations, for example, cloud services and GPUs. The control components are used to instrument and actuate the physical systems, for example, temperature adjustment and remote control. Communications provide the information and data transmission among IoT components, e.g., LoRa and ZigBee wireless communications. The hardware refers to the physical systems that underlie the IoT network, for example, the manufacturing plant and the robots.

The designer builds an IoT system using a blueprint $\delta : \mathcal{M} \mapsto \mathcal{D}$, which yields a design $d = \delta(m), d \in \mathcal{D}$ based on the device descriptions and specifications provided by the supplier. The system design leads to a performance $y \in \mathcal{Y}$. For example, in Example I, the designer develops a software system that integrates sensors, control algorithms, and the car. Safety is a critical performance measure of autonomous vehicles. It can be measured by the rate of accidents experienced by vehicles as of now. Here, we model the performance as a random variable. Given α and δ, the distribution of the performance random variable is $p_y(y; \theta, \alpha(m), \delta(m))$, $p_y : \Theta \times \mathcal{M} \mapsto \Delta \mathcal{Y}$. Using Bayes' rule, we arrive at

$$p_y(y; \theta, \alpha(m), \delta(m)) = p_y^\theta(y; \alpha(m), \delta(m)|\theta) p_\theta(\theta), \tag{4.1}$$

where $p_\theta(\cdot) \in \Delta\Theta$ is the prior distribution of the type of product; $p_y^\theta(y; \alpha(m), \delta(m)|$ $\theta)$, $p_y^\theta : \mathcal{M} \mapsto \Delta\mathcal{Y}$, is an indication of all possible system performances given the attribute of product θ. Note that the performance implicitly depends on m. The true performance of the system is determined by the true attribute of the product and the procurement and design decisions, which are made based on m. We denote $p_I = p_y(y; \theta, \alpha(\theta), \delta(\theta))$ as the ideal system performance when the design and procurement decisions are made given a truthful supplier, i.e., $m = \theta$.

Without knowing the true attributes of the product θ, the performance anticipated by the buyer is denoted by $q_y = p_y(y; m, \alpha(m), \delta(m))$. When $m \neq \theta$, there is a difference between the observed performance p_y and the anticipated one q_y. The buyer can perform hypothesis testing based on the sequence of observations $y_1, y_2, \cdots$, by setting up H_0 as the hypothesis that the observations follow the distribution q_y and H_1 otherwise. For example, in Example I, this decision is particularly important when y_i represents malfunctions or accidents for each trial test driving. If the malfunction is not expected by the designer, then there is a need to find out which supplier is accountable for the accidents or, in the case of a single supplier, whether the supplier should be held accountable.

4.3.3 Accountability Investigation

One critical step of accountability is the ability to attribute the performance outcomes to the supplier. We start with the accountability of a single supplier with binary type $\Theta = \{0, 1\}$ and assume the message space is the same as type space $\mathcal{M} = \Theta$. Consider a sequence of repeated but independent observations $Y^k = \{y_1, y_2, \cdots, y_k\}$, $k \in \mathbb{N}$. A binary accountability investigation is performed based on Y^k. Based on the received m, hypothesis H_0 is set to be the case when the observations follow the anticipated distribution q_y and H_1 otherwise. Depending on whether H_0 or H_1 holds, each observation y_i admits the following distribution

$$H_0 : y_i \sim f_m(y|H_0) = p_y(y; m, \alpha(m), \delta(m)), \tag{4.2}$$

$$H_1 : y_i \sim f_m(y|H_1) = p_y(y; \neg m, \alpha(m), \delta(m)). \tag{4.3}$$

The optimum Bayesian investigation rule is based on the likelihood ratio, which is denoted by

$$L(Y^k) = \prod_{j=1}^{k} \frac{p_y(y_j; \delta(m)|\neg m) p_\theta(\neg m)}{p_y(y_j; \delta(m)|m) p_\theta(m)}, \tag{4.4}$$

where we omit the purchase decision because the performance can only be observed when $a = 1$ and $\alpha(m) = Pr[a = 1|m]$ is the same under both hypotheses. The likelihood ratio test (LRT) provides the decision rule that H_1 is established when $L(Y^k)$ exceeds a defined threshold value $\tau_k \in \mathbb{R}$; otherwise, H_0 is established. It can be formulated by the equation

$$L(Y^k) \underset{H_0}{\overset{H_1}{\gtrless}} \tau_k. \tag{4.5}$$

One critical component in accountability investigation is the prior distribution over hypotheses, which indicates the reputation of the supplier. Without knowing the true distribution of the type, we argue that reputation is sufficient knowledge to determine the accountability of the supplier. Here, we give the definition of reputation over a binary type space, but the definition can be extended to multiple type space accordingly.

Definition 4.1 (Reputation) The reputation of the supplier $\pi \in \Delta\mathcal{H}$ is a prior distribution over all hypotheses. In binary case, $\pi_0 = \Pr[H_0]$ is the prior probability that the supplier truthfully report and $\pi_1 = \Pr[H_1]$ otherwise, with $\pi_0 + \pi_1 = 1$.

Assume that the cost of the investigation is symmetric and incurred only when an error occurs. In the binary case, the optimum decision rule will consequently minimize the error probability, and the threshold value τ_k in LRT will reduce to

$$\tau_k = \pi_0/\pi_1. \tag{4.6}$$

Definition 4.2 (Accountability)

1. Given an investigation rule, i.e., the threshold τ_k, the accountability $P_A \in [0, 1]$ is defined as the probability of correct establishment of hypothesis H_1 based on the observations Y^k and message m, which is given by

$$P_A(\tau_k) = \int_{\mathcal{Y}_1} f_m(Y^k|H_1) dy^k, \tag{4.7}$$

where $\mathcal{Y}_1$ is the observation space where $\mathcal{Y}_1 = \{Y^k : L(Y^k) \geq \tau_k\}$.
2. The wronged accountability $P_U \in [0, 1]$ is defined as the probability of a false alarm that H_1 is established while the underlying truth is H_0. Consider the threshold τ_k and observations Y^k, P_U is given by

$$P_U(\tau_k) = \int_{\mathcal{Y}_1} f_m(Y^k|H_0) dy^k. \tag{4.8}$$

We call a supplier η-unaccountable if $P_A \le \eta$, for a threshold accountability $\eta \in [0, 1]$ chosen by the investigator. In this case, the system does not have strong confidence that the observed accidents are caused by the supplier. We call a system ϵ-nontransparent if $P_A \le \epsilon$, for a given small $\epsilon \in [0, 1]$. That is, the system is close to being unable to hold the vendor accountable for the accidents.

The performance of the accountability investigation will be evaluated in terms of P_A and P_U. Ideally, we would like to conduct error-free accountability testing where P_A is close to one and P_U is close to zero (correctly identify accountable supplier without making mistake). However, the definition above leads to a fundamental limit on the accountability of the supplier. Except for situations where the observations Y^k under H_0 and H_1 are completely separable or the number of observations k goes to infinity, the performance of the accountability testing will be restricted within a feasible region.

> **Definition 4.3 (Accountability Receiver Operating Characteristic)** Accountability Receiver Operating Characteristic (AROC) is a plot which describes the relationship between achievable accountability P_A and wronged accountability P_U in the square $[0, 1] \times [0, 1]$.

As shown in Fig. 4.5, if we conduct LRT in accountability investigation, the AROC curve depicts the testing performance with respect to different threshold values τ_k. Similar to traditional binary hypothesis testing, the AROC curve under proper design preserves the following properties [70].

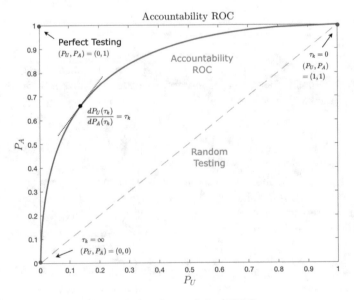

Fig. 4.5 Accountability receiver operating characteristics (AROC)

Property 4.1 (AROC)   AROC curve under proper design has the following properties:

(1) $(P_U, P_A) = (0, 0)$ and $(1, 1)$ belong to the AROC.
(2) The slope of the AROC curve $dP_A(\tau_k)/dP_U(\tau_k)$ is equal to the threshold τ_k.
(3) The AROC curve is concave and the feasible domain of (P_U, P_A) is convex.
(4) $P_A(\tau_k) \geq P_U(\tau_k), \forall \tau_k \in [0, +\infty)$.

Remark 4.1   The likelihood ratio lies in the region between zero and infinity. If we set the threshold τ_k in LRT to zero, investigator will classify any performance results into hypothesis H_1 (misinformation). Both accountability P_A and wronged accountability P_U will approach to one, as $(P_U, P_A) = (1, 1)$. Similarly, if we set τ_k in LRT to infinity, investigator will classify any performance into hypothesis H_0 (truthfully report), resulting in $(P_U, P_A) = (0, 0)$.

Remark 4.2   Property (3) and (4) are satisfied under the proper design; i.e., the test is "good" when $P_A \geq P_U$. For a "bad" test when $P_A < P_U$. As the hypothesis refers to a specific context of applications, we cannot simply reverse the performance distribution as in traditional hypothesis testing. Instead, we need to re-construct the investigation and find another performance metric that can properly distinguish the misinformation between suppliers and buyers.

It is worth noting that as the threshold τ_k increases, the accountability of the supplier P_A increases. However, according to the aforementioned properties, it would also increase wronged accountability P_U when the accidents are not caused by the vendor. There is a fundamental trade-off between accountability P_A and wronged accountability P_U depending on the accountability investigation. One way to evaluate the investigation performance is the area under the AROC curve (AUC). AUC is a measure of investigation capability [71], which provides a simple figure of merit to represent the degree of separability between two hypotheses.

$$AUC(\tau_k) = \int_0^1 P_A(\tau_k)\, dP_U(\tau_k) \tag{4.9}$$

This value varies from 0.5 to 1. When AUC equals 0.5, the designed investigation has no separation capability, which means the performance of the test is no better than flipping a coin. This corresponds to the case when $P_A(\tau_k) = P_U(\tau_k)$ for all possible threshold τ_k. Ideally, an excellent test will produce an AUC equal to one. In this situation, the accountability investigation can completely distinguish between two hypotheses, thus correctly identifying the supplier who should be accountable for the accidents.

Unfortunately, in realistic investigation tasks, it is hard to obtain the exact computation of AUC. Analyzing the upper and lower bounds of AUC helps the investigator describe the performance of the designed test. Shapiro in [72] provides an upper bound and lower bound on binary testing. Consider equally likely hypotheses with $\tau = 1$, the probability of error $P_e \in [0, 1]$ is defined as

$$P_e = \frac{P_U(\tau = 1)}{2} + \frac{1 - P_A(\tau = 1)}{2}. \tag{4.10}$$

Due to the convexity of the AROC curve, the bounds of the AUC can be described as

$$1 - P_e \leq AUC \leq 1 - 2P_e^2. \tag{4.11}$$

4.3.4 Model Extensions

This framework can be extended to multiple product types and multiple suppliers. Accountability needs to point to varied suppliers that cause failures under the hypothesis. In this section, we provide several testing frameworks and the definition of accountability accordingly.

4.3.4.1 Single Supplier with Multiple Types

Consider the product from the supplier with $T \in \mathbb{N}$ possible types, $\Theta = \{\theta_1, \theta_2, \ldots, \theta_T\}$. Based on the received message $m = \theta_m$, hypotheses $\{H_1, H_2, \ldots, H_T\}$ can be constructed by the investigator such that the performance observation y under each hypothesis H_t admits

$$H_t : y \sim f_m(y|H_t) = p_y(y; \theta_t, \alpha(\theta_m), \delta(\theta_m)), \tag{4.12}$$

for $1 \leq t \leq T$. The distribution under hypothesis H_t describes the system performance if the buyer makes purchase and designs based on the message θ_m while the underlying true product type is θ_t. In this case, the only anticipated performance by the buyer follows H_m. Any other observation distribution $H_{t \neq m}$ will attribute to the accountability of the supplier. Investigation could be conducted through M-ary hypothesis testing. For a single supplier with multiple product types, we can define the accountability as follows.

Definition 4.4 (Accountability with Multiple Types) Given a detection rule λ, received message m and observations Y^k, the accountability for a single supplier with multiple product types is defined as

$$P_A(\lambda) = \sum_{t \neq m, 1 \leq t \leq T} \int_{\mathcal{Y}_t} f_m(Y^k|H_t) dy^k, \tag{4.13}$$

where $\mathcal{Y}_t$ is the observation space we classify the observations as H_t.

If we assume that the investigation cost is symmetric and depends only on the error, it leads an MAP decision rule and the performance of the accountability testing can be evaluated through the error probability as

$$P_e = \sum_{1 \leq t \leq T} \Pr(E|H_t)\pi(t), \qquad (4.14)$$

where E denotes the error event, and $\pi(\cdot) \in \Delta\Theta$ is the prior probability that H_t is true, which represents the reputation of the supplier.

4.3.4.2 Multiple Suppliers

In IoT system design with multiple suppliers, accountability testing needs to point to varied suppliers that cause failures under the hypothesis. To simplify the illustration, we consider the case where the component from each supplier may have binary types $\theta_i \in \{0, 1\}, \forall i \in I$. Consider the problem with N vendors in the supply chain. Each supplier $i \in I = \{1, 2, \ldots, N\}$ with true product type θ_i will send a message $m_i \in M_i$ to the buyer to make purchase decision $a_i \in \{0, 1\}$ and determine the overall design $d \in D$. The process is illustrated in Fig. 4.6. We can construct hypotheses as a vector

$$H_j = (h_1, h_2, \ldots, h_N), \quad h_i = \mathbb{1}(m_i \neq \theta_i) \forall i \in I, \qquad (4.15)$$

where each element h_i is an indicator of whether supplier i truthfully reports or not, and the subscript $0 \leq j \leq 2^N - 1$ is the decimal number of the binary combination in the vector. The hypothesis vector indicates which supplier(s) should be accountable for the accident. When the performance distribution under each hypothesis is distinguishable, the investigation could be conducted through M-ary hypothesis testing. Otherwise, we can consider decentralized investigation as described in the sequel.

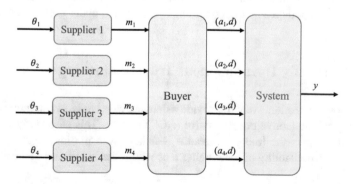

Fig. 4.6 Extension of the model to multiple suppliers

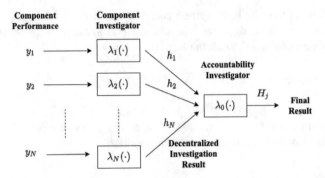

Fig. 4.7 Accountability by aggregating the investigations of the performance of multiple IoT components. A local investigation is performed based on the performance of each component $y_1, y_2, \cdots, y_N$. The results $h_1, \cdots, h_N$ are aggregated to form a final result H_j

Consider a decentralized accountability investigation with 2^N hypothesis $H_0, .., H_{2^N-1}$ and prior reputation $\pi(H_0), \ldots, \pi(H_{2^N-1})$, respectively. Suppose that we have N suppliers providing components to the system. Each component investigator λ_i is inspecting the performance related to the product provided from the vendor i. In practice, we can design the independent tests for each component to determine the accountability of supplier i. We can control the other parts $(j \neq i)$ to be known and fixed products in test design and focus on the binary hypothesis testing with respect to component i.

Illustrated in Fig. 4.7, each component investigator receives observations y_i, which is a random variable taking values in a set $\mathcal{Y}_i$. The local investigator will conduct accountability testing through $\lambda_i : \mathcal{Y}_i \mapsto \{0, 1\}$ and output a binary decision variable $h_i = \lambda_i(y_i)$, which indicates whether supplier i should be held accountable for the accident. This reduces the problem to N parallel binary hypothesis testing with each supplier, and the accountability of each supplier then will be the same as defined in Definition 4.2. The final investigator determines which hypothesis is established based on received information, $\lambda_0 : \{0, 1\}^N \to \{0, 1, \ldots, 2^N - 1\}$. It has been shown in [73] and [74] that there exists an optimal detection rule if each testing observations are independent or conditionally correlated under each hypothesis.

4.4   Case Study 1: Autonomous Truck Platooning

In the following section, we will provide a detailed case study in autonomous truck platooning with adaptive cruise control (ACC) system. This case study illustrates the scenario where the true performance is unknown to the investigator. We will discuss the accountability of the ranging sensor supplier in the case of a collision.

4.4.1   Background

With the rapid development of autonomous vehicles, safety is one of the main priorities for manufacturers. As estimated by the World Health Organization (WHO), the number of annual road deaths with collision has reached 1.35 million worldwide [75]. The recent incident in Tempe, Arizona, has thrown a spotlight on the safety of autonomous vehicles. The Uber self-driving test car caused the death of the pedestrian because of the failure of braking control by the autonomous driving system. The investigation of accountability is crucial to determine the cause of the collision and provides insights for future car design.

In this case study, we consider the task of autonomous truck platooning with ACC system. Adaptive cruise control is a driver assistance technology that maintains a safe following distance between the vehicle and traffic ahead without any intervention by the driver. If the preceding truck is detected traveling too slowly or too close, the ACC system will react by automatically activating the brakes and mitigating potential collisions. Brake control is determined based on the relative distance, relative velocity, and the acceleration of leading and the following truck. The speed and acceleration of both vehicles can be measured by built-in speed sensors and accelerometers. Ranging sensors, including radar and LiDAR, are used for distance detection in the ACC system. The upper-level control system uses the measurements of the sensors to interpret the driving environment, and trigger appropriate brake actions to mitigate collision [76]. Thus, the detection range and precision of the ranging sensor are critical in ACC design. Defective ranging sensors could cause severe consequences and should be held accountable in case of such a collision.

4.4.2   Vehicle Dynamics Model

To illustrate the accountability of the ranging sensor in this framework, we first introduce the dynamics model of the problem. Consider the testing scenario in Fig. 4.8, where the host truck equipped with ACC system approaches the preceding vehicle. The control goal of the ACC system is to maintain the desired safe distance from the leading vehicle. The desired distance L is normally determined by *Constant time gap* spacing policy in ACC systems, which guarantees the individual vehicle stability and string stability [76].

$$L = v_h \cdot t_{gap}, \tag{4.16}$$

where v_h is the speed of the host vehicle and t_{gap} is the constant desired time gap.

Denote x_i, v_i, a_i as the position, velocity, and acceleration of the leading $(i = l)$ or host $(i = h)$ vehicle, respectively. We assume that the leading vehicle

Fig. 4.8 Host truck with ACC system following the leading truck

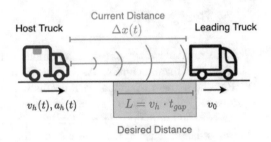

is at constant speed $v_l(t) = v_0$. The system state vector $\mathbf{x}(t)$ and control vector $\mathbf{u}(t)$ are defined as follows [77].

$$\mathbf{x}(t) = \left[\Delta x(t) - L, \ \Delta v(t)\right]^T, \quad \mathbf{u}(t) = \left[a_h(t)\right], \qquad (4.17)$$

where $\Delta x(t) = x_l(t) - x_h(t)$ is the current distance and $\Delta v(t) = v_l(t) - v_h(t)$ is the relative speed between the leading and following vehicles. The state space representation of the system can be written as

$$\dot{\mathbf{x}}(t) = A\mathbf{x}(t) + B\mathbf{u}(t), \qquad (4.18)$$

$$y(t) = C\mathbf{x}(t) + w(t), \qquad (4.19)$$

The matrices are given by

$$A = \begin{bmatrix} 0 & 1 \\ 0 & 0 \end{bmatrix}, \quad B = \begin{bmatrix} -t_{gap} \\ -1 \end{bmatrix}, \quad C = \begin{bmatrix} 1 & 0 \end{bmatrix}, \qquad (4.20)$$

where $y(t) = \Delta x(t) - L + w(t)$ is the noisy control error between the desired distance and current distance; $w(t)$ is the observation noise. We assume that the observation disturbance is modeled by an additive white Gaussian noise,

$$w(t) = \mathcal{N}(0, \sigma^2). \qquad (4.21)$$

The variance σ^2 indicates the influence of the measurement environment. The intuition behind using the Gaussian noise model is that it gives a good approximation of the natural processes. If a specific distribution of measurement error is given, the noise model can be changed accordingly and the accountability testing framework will still work.

The optimal control can be achieved through linear quadratic regulator (LQR) control. We define the cost function with zero terminal cost as

$$J = \frac{1}{2} \int_{t=0}^{\infty} \mathbf{x}(t)^T Q\mathbf{x}(t) + \mathbf{u}(t)^T R\mathbf{u}(t) \, dt, \qquad (4.22)$$

where the diagonal weights

$$Q = \begin{bmatrix} w_1 & 0 \\ 0 & w_2 \end{bmatrix}, \quad R = \begin{bmatrix} 1 \end{bmatrix}. \tag{4.23}$$

The goal of the controller is to regulate the state towards $(0, 0)^T$. The optimal feedback control law is given as

$$u(t) = -R^{-1} B^T P \mathbf{x}(t) \tag{4.24}$$

where P is the solution to the following associated algebraic Riccati equation:

$$0 = PA + A^T P + Q - PBR^{-1}B^T P. \tag{4.25}$$

The aforementioned vehicle dynamics model and optimal control describe the system design δ of the final ACC system based on the information provided by the supplier. Different control methods and system design can be implemented to achieve the same goal. In the following section, we assume that this system design is not the cause of the collision and purely focuses on the accountability of the sensor supplier.

4.4.3   Accountability Testing

The true product attributes play an important role in control system design. From the previous section, the optimal control of the system depends on the correct distance detection between the two objectives. Thus, the sensor with degraded detection result should hold accountable if the ACC system fails to maintain the safety distance and causes a collision. To attribute the ACC system performance to the ranging sensor supplier, we conduct the following accountability testing with respect to the ranging sensor.

For the simplicity of the model, we consider two types of ranging sensor $\theta \in \Theta = \{0, 1\}$, which differ in the detection precision. We assume the sensor with type $\theta = 1$ is functioning normally, as the detection result $r_1(t) = \Delta x(t)$; while the sensor with type $\theta = 0$ is malfunctioning with detection result $r_0(t) = \Delta x(t) + e_d$. The value e_d is the detection error of the ranging sensor. The damaged sensor will put the host vehicle at risk of collision, since the actual distance is closer to the detection result.

The true property of the sensor is private information to the supplier, which is not revealed to the system designer. The supplier should hold accountable for a collision if there exists misinformation between the product description m and true product property θ. Note that the misinformation can be unintended or intentional. We would like to determine whether the ranging sensor supplier should be accountable for such an accident.

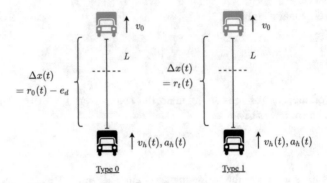

Fig. 4.9 Accountability testing with different sensor types

Consider the testing scenario in Fig. 4.9. The distance detection result from the sensor will be the input of the state vector as

$$\mathbf{x}(t) = \left[r_\theta(t) - L, \ \Delta v(t)\right]^T . \tag{4.26}$$

We use the final distance control error as the performance y of the ACC system when testing. Suppose that the supplier reports $m = 1$ when signing the contract. Consider a noisy observation results y as described in (4.19), then the performance should follow

$$y \sim p_y(y; 1, \alpha(1), \delta(1)) = N(0, \sigma^2).$$

It is the anticipated distribution of the observations when the supplier truthfully report the product type ($m = \theta = 1$). On the other hand, if the supplier misinforms the buyer ($m \neq \theta = 0$), the performance should follow

$$y \sim p_y(y; 0, \alpha(1), \delta(1)) = N(-e_d, \sigma^2)$$

The negative distance control error suggests that the distance between two vehicles is smaller than the desired safety distance requirement L, which can lead to a potential collision.

We set up the following hypotheses to quantify the accountability of the supplier who reports $m = 1$. Let $\mathbf{Y} = [y_1, y_2, \ldots, y_N] \in \mathbb{R}^N$ be a vector of independent identically distributed observations y_k ($1 \leq k \leq N$) of the aforementioned testing scenarios.

$$H_0 : \mathbf{Y} \sim N(-e_d, \sigma^2 \mathbf{I}_N)$$

$$H_1 : \mathbf{Y} \sim N(0, \sigma^2 \mathbf{I}_N)$$

where $\mathbf{I}_N$ is the identity matrix of size N. To keep the consistency with other studies, we let H_1 represent the case that the supplier truthfully report and H_0 mean that there exists misinformation between the reported product description m and true product type θ. The supplier is accountable if the investigator correctly determines that hypothesis H_0 should be established.

Assume that the cost of the decision is symmetric and incurred only when an error occurs. The reputation of the supplier follows $[\pi_0, \pi_1]$. In Bayesian binary hypothesis testing, LRT compares the likelihood ratio to threshold $\tau = \pi_0/\pi_1$. The result suggests that the hypothesis H_0 be established if the sample mean S is smaller than the testing threshold η, as shown in the following

$$S = \frac{1}{N} \sum_{i=1}^{N} y_i \underset{H_0}{\overset{H_1}{\gtrless}} \eta \tag{4.27}$$

where

$$\eta = \frac{e_d}{2} + \frac{\sigma^2 \ln(\tau)}{N e_d} \tag{4.28}$$

Given the decision rule and supplier's reputation ratio τ, the accountability and wronged accountability of the sensor supplier who reported $m = 1$ is

$$P_A(\tau) = \int_{y_0} f_1(y|H_0)dy = 1 - Q\left(\frac{d}{2} + \frac{\ln(\tau)}{d}\right) \tag{4.29}$$

$$P_U(\tau) = \int_{y_0} f_1(y|H_1)dy = Q\left(\frac{d}{2} - \frac{\ln(\tau)}{d}\right) \tag{4.30}$$

where $Q(x)$ is the Gaussian Q function and $d = N^{1/2}e_d/\sigma$ [70].

4.4.4 Parameter Analysis

The accountability of the sensor supplier helps the investigator determine whether the failure of the ACC system should be attributed to the sensor. Since the accountability depends on parameters such as sampling size N, environmental observation noise variance σ^2 and sensor range difference e_d. In this section, we discuss several numerical results under different cases.

Figure 4.10 depicts the influence of the number of tests N and sensor detection error e_d on the accountability. First, we notice that the $P_A \to 1$ and $P_U \to 0$ as the number of tests N increases. This phenomenon indicates more testing will produce a more accurate detection of the supplier's accountability. From Eq. (4.27), we note that the observation means S converges almost surely to the expected mean of each

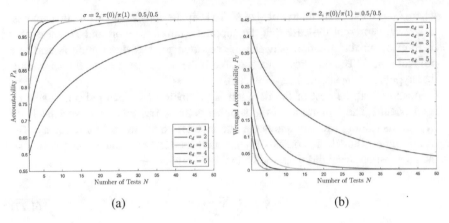

Fig. 4.10 Different sensor range difference ($\sigma = 2$, $\pi_0/\pi_1 = 0.5/0.5$). (**a**) Accountability P_A. (**b**) Wronged accountability P_U

Fig. 4.11 Impact of supplier's reputation ($\sigma = 2$, $e_d = 2$, $N = 30$)

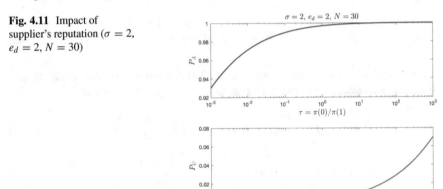

hypothesis as $N \to \infty$. Besides, the second term in the testing threshold η vanishes, and we end up comparing the expected mean of Y to the middle point $e_d/2$ of two hypothesis means.

The influences of sensor detection error e_d is also illustrated in Fig. 4.10. The prior is set to $\pi_0 = \pi_1 = 0.5$, which means that we do not favor any hypothesis before testing. From Fig. 4.10, as the range difference between two types increases, the P_A and P_D curves are associated with a more rapid change with respect to N. It suggests that if the qualities of the two types of sensors have a significant difference, it be easier for the investigator to determine the accountability of the supplier within a fewer number of tests.

Figure 4.11 displays the impact of supplier's reputation on the accountability estimation. The ratio $\tau = \pi_0/\pi_1$ represents the reputation of the supplier. A larger value of τ indicates that we have a strong belief the supplier is dishonest. Normally, we incline to expect that the supplier with a bad reputation would be accountable for the incidents. As shown in Fig. 4.11, when we fix the testing environment, the accountability of supplier P_A increases as τ increases. However, it should be noted that the wronged accountability P_U increases as well. This is because the increase of τ will cause the testing threshold η in LTR will increase, leading to a larger observation space $\mathcal{Y}_0$, where we classify the observations as H_0. Thus, both P_A and P_U will increase according to the definition. The wronged accountability misattributes the incident to the supplier when they should not be accountable. We will discuss the trade-off between P_A and P_U in the following section.

4.4.5 Investigation Performance

4.4.5.1 Accountability Receiver Operating Characteristic

In the context of this ACC case study, we are interested in the relationship between accountability P_A and wronged accountability P_U. as

$$P_A = \int_{\mathcal{Y}_0} f_m(\mathbf{y}|H_0)d\mathbf{y} = 1 - P_F \tag{4.31}$$

$$P_U = \int_{\mathcal{Y}_0} f_m(\mathbf{y}|H_1)d\mathbf{y} = 1 - P_D \tag{4.32}$$

Because of the symmetric property of the Gaussian Q function, the AROC curve is invariant under this transformation. From Eqs. (4.29) and (4.30), if we eliminate the parameter τ, the relationship between P_A and P_U can be written as

$$P_U = Q(d - Q^{-1}(1 - P_A)) \tag{4.33}$$

The relationship between P_A and P_U is traced out as the threshold τ in LRT varies from 0 to ∞. Note that this relationship depends on the variable $d = N^{1/2}e_d/\sigma$. We plot the AROC curve under different d values in Fig. 4.12.

The slope of the AROC at point $(P_A(\tau), P_U(\tau))$ is equal to the supplier's reputation τ[70]. Ideally, we would like to conduct a hypothesis test such that P_A is close to one and P_U is close to zero. As we can see from the figure, the AROC curve approached the ideal test point when the value of d increases. This result coincides with our aforementioned analyses. Increasing the number of test N, comparing sensor with larger sensor error e_d, and reducing the observation variance σ can all increase the value of d, leading to a more reliable accountability test result.

4.4.5.2 Area Under the AROC Curve

In the ACC sensor accountability testing case, the exact AUC value and its bounds with respect to d are shown in Fig. 4.13. From the figure, we can see that the

Fig. 4.12 ROC curve under different d

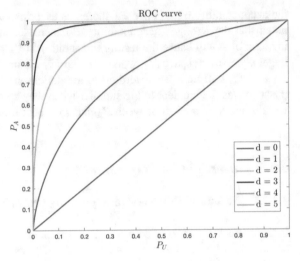

Fig. 4.13 Bounds of AUC under different d

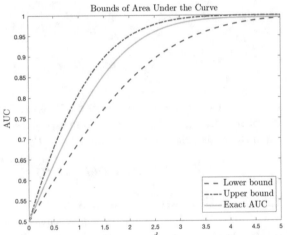

performance of the hypothesis testing increases along with the value d. In fact, in testing with the Gaussian hypothesis, the value d indicates the Chernoff distance between the two Gaussian distributions [70]. A larger value of d means the distribution of H_0 and H_1 have less overlap, thus it is easier to separate between them. Since we have the exact expression of P_e, the bounds of AUC can be expressed as

$$1 - Q\left(\frac{d}{2}\right) \leq AUC(d) \leq 1 - 2Q^2\left(\frac{d}{2}\right). \tag{4.34}$$

4.5   Case Study 2: Ransomware in IoT Supply Chain

In this section, we provide a second case study of supplier accountability in smart home IoT under ransomware attacks. This example illustrates how we determine accountability in a supply chain and sophisticated systems involving varied components.

4.5.1   Background

Ransomware is a type of malware that infects particular network entities to demand ransom. This kind of attack is becoming more prevalent nowadays with the fast development of IoT systems. The broad connections for IoT devices provide more security threats and vulnerabilities. Besides, the massive number of IoT devices increases the risk of getting infected by ransomware since any device could be the target. Indeed, the ransomware attack has caused significant economic losses in industrial domains. The estimated global damage from ransomware reaches $20 billion in 2021 [78].

Smart home technologies integrate different IoT-enabled components to provide advanced services within the home environment. The components from different suppliers contribute to addressing various challenges to improve the quality of human life. However, their limited processing capabilities make them vulnerable to security threats [79], including ransomware. If the component in the home security system is taken controlled by the attacker, the end-user may face serious economic loss and privacy leakage . The user needs to determine which part of the IoT system should hold accountable for the accident. Our framework provides a way to mitigate the impact of ransomware by attributing the accident to a supplier of IoT devices.

4.5.2   Smart Lock and Ransomware Attack

Nowadays, smart home technologies have been widely accepted by individuals and organizations to improve home security. With the development of IoT and machine learning, the number of smart lock users are increasing in recent decades. Instead of physical keys, smart lock utilizes face recognition and/or fingerprint verification to achieve digital authentication. Most smart locks also are equipped with intruder alert and remote control when you are physically away from home. This innovation avoids the threats with cloneable physical keys and provides a front-line deterrent against potential intruders.

While the smart lock offers convenience to homeowners, the transition towards digital control brings concerns over security in cyberspace. One potential threat is the ransomware attack. This type of attacks belongs to the family of Advanced Persistent Threats (APTs) . A malicious attacker attack your smart home IoT system,

Fig. 4.14 IoT supply chain related to security lock

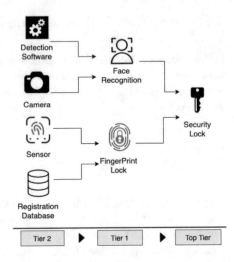

lock the front door of your house, and request a ransom. The highly-connected feature of IoT provides the attacker multiple vulnerabilities as the entry point into the network. Once building a foothold in the network, the attacker moves laterally towards the target to achieve his goal, in this case, locking the door and denying legitimate access. Once compromised by ransomware, the dangling participle would be huge if someone under medical conditions is locked and requires immediate treatment. We may be discouraged by the fact that victims simply pay the ransom in many cases, and even the FBI has inadvertently mentioned paying ransom if the network device is infected [80].

Accountability investigation provides a way to check the responsibility of the IoT device supplier(s) regarding the attack to mitigate the loss under such ransomware attacks. It is important for the investigator to find out the initial attack entry that poses a risk to the whole system. Due to the tiered structure of the supply chain, the accountability investigation needs to be constructed through a top-down layered tree analysis as shown in Fig. 4.14. This structure helps the investigator narrow down the search scope and determine the accountability of the suppliers among multiple supply chain tiers. More details are provided in the following section.

4.5.3 Accountability Investigation

4.5.3.1 Tier-1 Investigation

Face recognition and fingerprint verification are two critical parts of smart lock authentication. The failure of the smart lock could be caused by the failure of one or both of the functions. In this case, the first step in accountability investigation is to determine whether the tier-1 suppliers of these two parts need to be accountable for the ransomware attack.

Table 4.1 Four hypotheses
in accountability investigation

Hypothesis	$h_1 = \mathbb{1}(\theta_1 \neq 0)$	$h_2 = \mathbb{1}(\theta_2 \neq 0)$
$\hat{H}_0$	0	0
$\hat{H}_1$	0	1
$\hat{H}_2$	1	0
$\hat{H}_3$	1	1

Fig. 4.15 Decentralized
tier-1 accountability
investigation

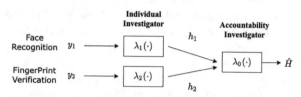

Denote the supplier of face recognition technology as $i = 1$ and the supplier of fingerprint verification technique as $i = 2$. We assume that each supplier has binary types $\theta_i \in \{0, 1\}$. $\theta_i = 0$ means that the provided product operates normally and $\theta_i = 1$ stands for malfunctioning. By default, each supplier sends a message $m_i = 0$ and guarantees the product functionality when signing the contract with the buyer. Thus, we can construct the following hypotheses as in Table 4.1. Denote $h_i, i = \{1, 2\}$, as the accountability of supplier i. $\hat{H}_0$ indicates that both parts operate normally as reported; $\hat{H}_1/\hat{H}_2$ suggests that there be misinformation from one of the suppliers; $\hat{H}_3$ means both suppliers need to be held accountable for the ransomware attack.

Instead of looking into the joint performance of the two components, we conduct independent decentralized investigations into each of the suppliers as shown in Fig. 4.15. We take the face recognition system h_1 for example. The investigation of the fingerprint verification h_2 can be conducted in the same manner. Suppose that the normal operating face recognition system can correctly detect the registered identity with $\mu_0 = 9\%$ accuracy. If this system is destructed by the ransomware attacker, we would expect a lower identification accuracy, i.e. $\mu_1 < \mu_0$. To investigate the accountability of the face recognition system, we design the following testing scenarios. On each trial, different photos of registered faces are displayed randomly in front of the device. The performance $y_i \in \{0, 1\}$ at each trial is an indicator of the testing results, where $y_i = 1$ represents correct identification and $y_i = 0$ otherwise. Let $Y^N = \{y_1, y_2, \ldots, y_N\}$ be a sequence of independent and identically distributed trials, we consider the following hypotheses for accountability testing. For each trail $1 \leq i \leq N$,

$$H_0 : y_i \sim \text{Bern}(0.9), \qquad H_1 : y_i \sim \text{Bern}(\mu_1),$$

where $\mu_1 < \mu_0 = 0.95$. Bernoulli distribution is a natural model to describe events with Boolean-valued outcomes under certain success probability. In this hypothesis model, H_0 indicates that the face recognition system operates normally with 90% detection accuracy on average. H_1 suggests a degraded identification accuracy. This investigation aims to find out whether hypothesis H_1 should be established based on the system performance.

One limitation of Bayesian tests as described in Sect. 4.4 is their reliance on the prior knowledge π, i.e., the reputation of the supplier, and costs assigned to different decision errors. The choice of decision cost depends on the nature of the problem, but the prior probabilities must be known. In many applications, the prior knowledge may not be obtained precisely; thus, the correct value of the threshold in LRT is unknown. In the ransomware case study, the misinformation between the supplier and buyer may be unintended. It is challenging to determine the probability π_1 that the supplier is compromised by the attacker. It is natural to consider alternative tests that can achieve desired detection results without such prior knowledge.

Neyman and Pearson [81] formulated a test λ that maximizes the correct detection probability $P_A(\lambda)$ (accountability) while ensuring the false-alarm probability $P_U(\lambda)$ (wronged accountability) is subject to an upper bound constraint α. This can be formulated as

$$\max_{\lambda} \quad P_A(\lambda) = \int_{\mathcal{Y}_1} f_m(Y^N|H_0)dy^N,$$

$$\text{s.t.} \quad P_U(\lambda) = \int_{\mathcal{Y}_1} f_m(Y^N|H_1)dy^N \leq \alpha. \tag{4.35}$$

This constrained optimization problem requires no prior knowledge about reputation and decision cost function. The only parameter that needs specification is the maximum acceptable wronged accountability α. A classic result due to Neyman and Pearson shows that the optimal solution to this type of investigation is a likelihood ratio test (LRT).

Lemma 4.1 (Neyman-Pearson Lemma) *Consider the likelihood ratio test in (4.5) with $\tau_k > 0$ chosen so that $P_U(\tau_k) = \alpha$. There does not exist another test λ such that $P_U(\lambda) \leq \alpha$ and $P_A(\lambda) \geq P_A(\tau_k)$. Hence, the LRT is the most powerful test with false-alarm probability $P_U(\lambda)$ less than or equal to α.*

In the accountability investigation of the face recognition system, both hypotheses admit a Bernoulli distribution. The likelihood ratio is given by

$$L(Y^k) = \frac{\prod_{i=1}^{N} \mu_1^{y_i}(1-\mu_1)^{1-y_i}}{\prod_{i=1}^{N} \mu_0^{y_i}(1-\mu_0)^{1-y_i}} = \left(\frac{1-\mu_0}{1-\mu_1}\right)^N \left(\frac{\mu_0(1-\mu_1)}{\mu_1(1-\mu_0)}\right)^{\sum_{i=1}^{N} y_i}.$$

The sufficient statistics of such testing will be the sum of all performance results $S = \sum_{i=1}^{N} y_i$. According to Neyman-Pearson lemma , the most powerful test will hold the supplier accountable if $S < \lambda$ for a constant threshold λ.

$$S = \sum_{i=1}^{N} y_i \underset{H_1}{\overset{H_0}{\gtrless}} \lambda$$

Fig. 4.16 Neyman-Pearson test result for tier-1 investigation

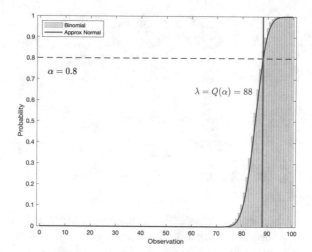

Under H_0, the detection accuracy is on average μ_0, and S admits to a binomial distribution, $S \sim \text{Binomial}(N, \mu_0)$. Illustrated in Fig. 4.16, to ensure $P_U(\lambda) = \alpha$, the threshold λ is chosen to be the α-quantile of the $\text{Binomial}(N, \mu_0)$ distribution.

$$\lambda = Q(\alpha) = \inf\{x \in \mathbb{R} : \alpha \leq F_S(x)\},$$

where $F_S(x)$ is the cumulative distribution function of random variable S. Note that as this is a discrete distribution, it may not be possible to get the exact α and λ desired. One way to address this problem is to increase the total number of trials N and approximate the binomial with a Gaussian distribution according to the central limit theorem (Fig. 4.16).

In the IoT ransomware attack case, the changes made by the stealthy attacker often remains unknown even after investigations. Thus, it is hard to determine identification accuracy μ_1 after the attack and find the exact performance distribution under hypothesis H_1. We can only assume that the attack results in a degraded identification accuracy as $\mu_1 < \mu_0$. Neyman-Pearson test provides a way to investigate the accountability of the supplier with limited prior knowledge. It guarantees that the correct detection probability P_A is maximized under the false-alarm constraint $P_U \leq \alpha$. In the context of the IoT supply chain attack, Neyman-Pearson test paves the way for the buyer to investigate the accountability of the supplier with limited information.

4.5.3.2   Multi-Stage Accountability Investigation

The tier-1 investigation examines the accountability of each tier-1 supplier. However, due to the layered structure of the IoT supply chain and the sophisticated feature of the ransomware attack, the true cause of the attack may lie in the suppliers

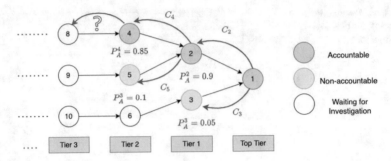

Fig. 4.17 Multi-stage accountability investigation

at the subordinate tiers. Tier-1 suppliers can further attribute the malfunction to their suppliers following a similar fashion. A top-down layered investigation is needed if we would find out the origin of the attack and obtain a holistic view of the entire supply chain. This is called a multi-stage accountability investigation.

For instance, if the face recognition system should be held accountable for the attack after the tier-1 investigation, the supplier could further investigate the components that the system consists of. There may exist different types of vulnerabilities in the components that are provided by tier-2 suppliers. The attacker could break into the system by compromising the ill-protected camera and further penetrating into the system. Another possibility is that adversaries against face recognition are performed at the detection software. If the latter case holds true, the detection software provider can further check which part of the software is malfunctioning. Face recognition attacks can be performed at the database, the predefined algorithm parameters, the communication channels, etc. The multi-stage accountability investigation aims to further figure out which among the vulnerabilities is the underlying cause of the attack (Fig. 4.17).

To analyze the accountability of the involved suppliers at each tier, we view the supply chain as a directed graph as shown in Fig. 4.19. The arrows in the graph indicate the procurement relationship. Multi-stage accountability starts from the top tier node, the final product. The accountability investigation on each supplier i produces accountability P_A^i subject to an investigation cost C_i. Whether a supplier is accountable depends on the comparison between P_A^i and selected threshold $\epsilon \in (0, 1)$. We call a supplier accountable if $P_A^i > \epsilon$.

If the current supplier is determined to be non-accountable ($P_A^i < \epsilon$), there is no need to continue investigation among its suppliers. In the ransomware case study, if we determine that the face recognition system solely should be accountable after the tier-1 investigation, there is no need to conduct an accountability check for the suppliers related to the fingerprint verification system. Deductive reasoning helps reduce the investigation efforts on unrelated system components and focus on the ones that the accident is attributed to. It provides a way to prioritize the factors leading to the top event.

It should be noted that the product design of each sub-system can also be the cause of the vulnerability that exposes the system to threats. This brings up the question of how deep we should investigate during the process. Suppose the total investigation budget is B. The investigator needs to decide whether to continue the investigation or simply stop and replace the component. Replacement is a better choice if the remaining budget cannot support further investigation as

$$B - \sum_{i \in \mathcal{I}} C_i \leq C_{next},$$

where $\mathcal{I}$ is the set of investigated suppliers and C_{next} is the investigation cost of the next supplier. The trade-off between investigation and replacement can be another dimension for consideration when conducting multi-stage accountability investigations.

Multi-stage accountability investigation is an iterative analysis process to find the cause of the accident. The layered approach provides a way to understand how the system fails, identify the vulnerabilities in the IoT supply chain, and determine the accountability of any supplier. It also creates the foundation for any further analysis and evaluation. If the structure of the supply chain has been upgraded (e.g., component replacement), it can provide a set of steps to design quality tests and maintenance procedures.

4.6 Compliance and Cyber Insurance

4.6.1 Compliance Modeling

The description $m \in M$ from the supplier to the buyer is a self-reporting mechanism that requires the vendors to disclose information about their products so that the buyers can use the NIST standards to check their compliance before they are integrated into IoT systems. The procured products have to comply with the business or mission, organization-specific requirements, the operational environment, risk appetite, and risk tolerance [82]. Security requirements are an important component of compliance. They are imposed by not only the developers in the private sectors to provide information and quality assurance but also the law, which aims to protect the nation from cyber-attacks.

Recent legislation has been signed into law requiring IoT devices purchased with government money to comply with security standards [83]. The Internet of Things Cybersecurity Act of 2020 [84] requires NIST to "develop and publish under section 20 of NIST Act (15 U.S.C. 278g-3) standards and guidelines for the federal government on the appropriate use and management of Internet of Things devices owned or controlled by an agency and connected to information systems owned or controlled by an agency, including minimum information security requirements for managing cybersecurity risks associated with such device." All IoT devices connected to IT systems owned or controlled by a federal agency must conform to NIST standards by September 4, 2021.

The Biden executive order of May 12, 2021 [85] demands that "the federal government must bring to bear the full scope of its authorities and resources to protect and secure its computer systems, whether they are cloud-based, on-premises, or hybrid. The scope of protection and security must include systems that process data (information technology (IT)) and those that run the vital machinery that ensures our safety (operational technology (OT))." The executive order requires full NIST compliance. The focus of the new rules is on IoT systems that support information technologies, e.g., the power and cooling systems, such as uninterruptible power supplies (UPSs), power distribution units (PDUs), and computer room air conditioners and air handlers (CRAC & CRAH) that support networks, servers, and data centers on the property of federal agencies, building management systems (BMS), and data center infrastructure management systems (DCIM).

Besides the federal regulations, supply contracts are also useful to secure systems installed by suppliers. The suppliers need to be informed of your security requirements and standards. You can check whether the proposed or delivered products or services comply with them. The contracts also play an important role in accountability. The penalty can be enforced by contracts once non-compliance of the services is found by the buyer, which has been discussed in the earlier section.

We can use formal methods to check whether the attributes in m satisfy the requirements that are coded into logical formulae f. The product is compliant if $m \models p$, the description satisfies the specifications; otherwise, it is not. There are well-established tools that can be used to efficiently solve this satisfiability problem. For example, the compliance problem can be formulated as a satisfiability modulo theories (SMT) problem, which can be solved using a formalized approach and many solvers. PRISM is another tool that enables probabilistic modeling and checking of systems. Under the assumption that the reporting of m truthfully describes the product, i.e., $m = \theta$, a compliant buyer or system will not acquire from suppliers that do not satisfy the requirement. In other words, $a = 0$ if $m \not\models p$.

4.6.2   Contract Design

There are two economic-level solutions. One is the mechanism design between the buyer and the supplier to induce $m = \theta$. To achieve this, we would need to create incentives for the supplier to truthfully reveal θ. This would rely on the design of a certain form of penalty as a credible threat. One of such penalties is through the contract. The contract between the supplier and the buyer would include a penalty once the supplier is accountable. The contract will be effective only when the buyer decides to purchase the product $a = 1$, which happens with probability $\alpha(m) = Pr(a = 1|m)$. We consider the following utility function of the supplier, $U_S : \Theta \times \mathcal{M} \mapsto \mathbb{R}$, given by

$$U_S(\theta, m) := \mathbb{E}_\alpha \left[J_S(\theta, m) - \mathbb{E}_{P_A^m} [C_S(\theta, m)] \right]. \tag{4.36}$$

Here, $J_S : \Theta \times \mathcal{M} \mapsto \mathbb{R}$ is the profit of the supplier if he reports $m \in \mathcal{M}$ when the true type is $\theta \in \Theta$ and under the procurement decision. The second term in the utility function is the average penalty $C_S : \Theta \times \mathcal{M} \mapsto \mathbb{R}$ for the supplier if he is held accountable. The probability of being accountable is given by P_A^m in Definition 4.2 based on the received message m. It is clear that the penalty depends on θ and m.

We call a supplier is incentive-compatible if

$$U_S(\theta, \theta) \ge U_S(\theta, m), \quad \text{for all } m \in \mathcal{M}. \tag{IC_S}$$

An incentive-compatible supplier does not have incentives to misreport what he knows when he is held accountable for his actions. Note that to achieve this, we assume that the purchase rule and accountability testing scheme are revealed to the supplier through the contract. The (IC_S) condition gives a natural constraint when designing a procurement contract . However, the challenge is that the profit function J_S and the type space of the suppliers are often unknown to the acquirer and they need to be conjectured or learned from experience or data.

We call a supplier is individually rational if

$$U_S(\theta, m) \ge 0, \quad \text{for all } m \in \mathcal{M}, m \ne \theta \tag{IR_S}$$

The (IR_S) constraint ensures that the supplier benefits from participating in the contract. It requires the buyer to design the penalty carefully so that the expected profit of the supplier is non-negative.

Example: Autonomous Truck Platooning

The utility function of the supplier can be further expressed as

$$U_S(\theta, m) = \alpha(m) \cdot \left[J_S(\theta, m) - C_S(\theta, m) \cdot P_A^m \right]. \tag{4.37}$$

The goal of contract design is to assign an appropriate penalty C_S for the supplier if they need to be held accountable for the accident. The first consideration comes from the (IR_S) constraints. This set of constraints suggests that we should not assign a penalty that exceeds the expected profit.

The (IC_S) constraints are automatically satisfied when the supplier truthfully reports $m = \theta$. Consider the autonomous truck platooning example as described in Sect. 4.4 with the binary sensor type space, i.e., $\Theta = \mathcal{M} = \{0, 1\}$. The contract designer needs to meet the following constraints

$$\alpha(1) \left(J_S^{11} - P_A^1 C_S^{11} \right) \ge \alpha(0) \left(J_S^{10} - P_A^0 C_S^{10} \right) \tag{4.38}$$

$$\alpha(0) \left(J_S^{00} - P_A^0 C_S^{00} \right) \ge \alpha(1) \left(J_S^{01} - P_A^1 C_S^{01} \right) \tag{4.39}$$

where we denote the profit of supplier with true type θ who sends message m as $J_S^{\theta,m}$, and the penalty for such supplier as $C_S^{\theta,m}$.

From the contract designer's viewpoint, the profit of the supplier $J_S^{\theta,m}$ is beyond his control. This value is determined by the production cost and economic nature of the system. In the ACC system, $\theta = 1$ is the product type corresponding to the system design. It is natural to assume that the sensor supplier with true type $\theta = 1$ makes more profit when he truthfully reports, as $J_S^{11} > J_S^{10}$. Similarly, we can assume that misinformation brings a higher profit for the supplier with $\theta = 0$, as $J_S^{00} < J_S^{01}$.

In terms of misinformation penalty, it is incentive to penalize more on the supplier who fails to truthfully report, as $C_S^{\theta,\theta} < C_S^{\theta,m}$, for every $m \neq \theta$. If we expect the same procurement policy $\alpha(m)$ and accountability $P_A^m = P_A$ are the same for both messages $m \in \{0, 1\}$, constraint (4.38) will be automatically satisfied and constraint (4.39) will be reduced to

$$J_S^{01} - J_S^{00} \leq P_A(C_S^{01} - C_S^{00}). \tag{4.40}$$

This indicates for the supplier $\theta = 0$ who has the incentive to misinform the buyer, the expected extra penalties brings to the supplier through contract need to exceed the extra profit generated from the untruthful report. The result coincides with the intuition that the contract needs to be designed with incentive compatibility.

For automakers looking at production, the prices of LiDAR sensors need to be cost-effective for automotive ACC use. Ranging sensors with greater abilities will be sold for higher prices. It is reported that LiDAR suppliers manage to reduce the single-unit samples price to \$250 in large volumes [86]. In the ACC supplier example, consider the following values:

$$J_S^{11} = J_S^{01} = 250; \quad J_S^{00} = J_S^{10} = 200; \quad \alpha(1) = 0.8, \alpha(0) = 0.5; \quad P_A^1 = 0.3, P_A^0 = 0.7.$$

We arrive at the following constraints for the contract penalty design for the supplier:

$$0.8 * (250 - 0.3 * C_S^{11}) \geq 0.5 * (200 - 0.7 * C_S^{10}),$$

$$0.5 * (200 - 0.7 * C_S^{00}) \geq 0.8 * (250 - 0.3 * C_S^{01}), \qquad (IC_S)$$

$$0.5 * (200 - 0.7 * C_S^{10}) \geq 0,$$

$$0.8 * (250 - 0.3 * C_S^{01}) \geq 0, \qquad (IR_S)$$

$$C_S^{00} < C_S^{01}, \quad C_S^{11} < C_S^{10}.$$

By solving the feasible region of penalty under constraints as in Fig. 4.18, the contract designer can select the proper penalties for the supplier and help avoid misinformation.

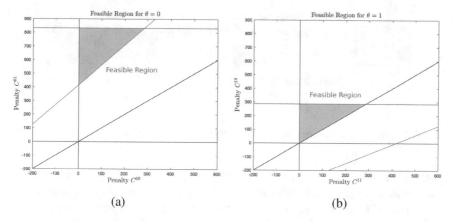

Fig. 4.18 Feasible penalties under constraints. (**a**) Feasible region for $\theta = 0$. (**b**) Feasible region for $\theta = 1$

4.6.3 Cyber Insurance

4.6.3.1 Background Introduction

In spite of the wide applications of cyber-physical systems, the cyber risks within the IoT supply chain are considered to be the most challenging problem to handle. Cyber insurance is the last resort for resilience to mitigate the loss of performance. It is an important risk management tool that transfers the risks of the buyer to a third party, i.e., an insurer. Victims of a cyber attack can reduce their financial losses and quickly recover to restore their business operations. According to the cyber insurance report released by the National Association of Insurance Commissioners (NAIC) [87], the cybersecurity insurance market in 2020 is roughly $4.1 billion reflecting an increase of 29.1% from the prior year. This scheme particularly benefits small and medium-size businesses that cannot afford a major investment in cyber protection.

Unlike traditional insurance policies, cyber insurance compensates the buyer for the loss incurred by data breaches, malware infections, or other cyberattacks in which the insured entity was at fault. An incentive-compatible cyber insurance policy could help reduce the number of successful cyber attacks by incentivizing the adoption of preventative measures in return for more coverage [88, 89]. It can be served as an indicator of the quality of security protection. Besides, it is believed that cyber insurance can induce greater social welfare and encourage more comprehensive policies regarding cyber security[90].

Various frameworks have been proposed to study cyber insurance from different perspectives, including [17, 65, 91, 92]. Pal et al. have studied the economic impact of cyber insurance by proposing a supply-demand model. Their work showed that cyber insurance with client contract discrimination can improve network security [93]. Böhme et al. have proposed several market models to understand

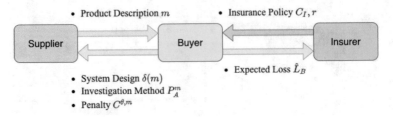

Fig. 4.19 Information exchange between the insurer, buyer and supplier

the information asymmetries between defenders and insurers [94]. Radanliev et al. have built an impact assessment model of IoT cyber risk to better estimate cyber insurance [95]. In our framework, we focus on the cyber insurance policy within the IoT supply chain and understand the impact of accountability investigation on cyber insurance.

4.6.3.2 Insurance Policy Design

Typically, the cyber insurance contract consists of the premium price and the coverage rate . The key challenge in insurance policy design lies in the difficulty of risk evaluation due to the complex structure of the cyber-physical systems. An insurer can make two separate contracts with the supplier or/and the buyer. The loss of the buyer would be compensated by the insurer when an accident or a disruption occurs. The loss of the supplier due to accountability could be insured as well. In this section, we focus on the insurance contract between an insurer and a buyer (Fig. 4.19).

The contract is composed of the premium and the coverage of the losses. Let $C_I \in \mathbb{R}$ be the premium charged by an insurer and the coverage is modeled by the percentage $r \in (0, 1]$. They are decision variables that are determined by the insurer. A buyer has incentives to participate in the insurance if the average utility under the coverage is higher than the one without coverage. To quantitatively capture it, we specify the loss or payoff function of the buyer J_B, given by

$$J_B(m, \delta) := (1 - r)\hat{L}_B(m, \delta(m); \theta) + C_B(m) + C_I. \tag{4.41}$$

Here, the first term $\hat{L}_B$ is the average loss of performance, which is the difference between the true and the anticipated performances. The cyber insurance will cover the r portion of the risk. Hence the residual loss is $(1-r)$ of the losses. The insurance can completely compensate for the loss of the performance when $r = 1$. The second term is $C_B(m)$ is the cost of procurement of the product and C_I is the premium paid by the buyer.

In this framework, we focus on the potential loss due to the misinformation from the supplier who cannot be held accountable due to the limitation of accountability investigation. According to the investigation, if the supplier should be held accountable for the malfunctioning of the system, the loss of performance should be compensated by the supplier. However, if the investigation cannot hold the supplier accountable, the risk will be transferred to the third party under the insurance contract. The latter case occurs with probability $1 - P_A^m$, *the probability of unaccountable*. Thus, the loss of performance can be viewed as a random variable l_B

$$l_B(m, \delta(m); \theta) = \begin{cases} U_B(m, \delta(m)) - U_B(\theta, \delta(m)) & \text{w.p. } 1 - P_A^m, \\ 0 & \text{w.p. } P_A^m, \end{cases} \tag{4.42}$$

where $U_B(\theta, \delta(m))$ is the performance utility measure under the design $\delta(m)$ and the true product quality θ. We assume that the true performance $U_B(\theta, \delta(m))$ is at best the same as the anticipated performance when $m = \theta$, i.e. $U_B(m, \delta(m))$. When misinformation occurs, there will be a positive loss of performance; when the supplier reports truthfully, the true performance coincides with anticipated one and the loss is zero; in other words, the expected loss of performance

$$\hat{L}_B = (1 - P_A^m)\Delta U_B \geq 0, \tag{4.43}$$

where we denote the difference in performance measure as ΔU_B.

One critical aspect of cyber insurance is the bias from insurance buyers. Humans will hold biased perception concerning losses and risks, which can lead to different decisions compared to completely rational ones. Agents are often risk-averse; i.e., they prefer lower returns with known risks rather than higher returns with unknown risks. In terms of the expected losses $\hat{L}_B$, economic literature commonly imposes the following functions for a risk-averse agent.

- Constant Absolute Risk Aversion (CARA) [94]:

$$\phi(x) = \frac{e^{\beta x}}{\beta}, \tag{4.44}$$

where the parameter $\beta \leq 1$ is the absolute risk aversion coefficient, measuring the degree of risk aversion that is implicit in the utility function. The biased expected loss in this case is

$$\Phi(\hat{L}_B) = (1 - P_A^m)\phi(\Delta U_B), \tag{4.45}$$

- Prospect Theory (PT) [96]:

$$\phi(x) = \begin{cases} x^\beta & x \geq 0 \\ -\lambda(-x)^\beta & x < 0 \end{cases}, \quad w(p) = \frac{p^\zeta}{p^\zeta + (1-p)^\zeta}, \tag{4.46}$$

where $\phi(x)$ and $w(p)$ are biased utility and weighted probability, respectively, and λ, β, ζ are prospect parameters with loss aversion implying $\lambda > 1$. In general, PT shows that people are more averse to losses and less sensitive to gains; people inflate the belief for rare events and deflate for high-probability ones. The biased expected loss in this case is

$$\Phi(\hat{L}_B) = w(1 - P_A^m)\phi(\Delta U_B),\tag{4.47}$$

For these types of buyer, we should replace the average loss $\hat{L}_B$ in Eq. (4.41) with the biased expectation $\Phi(\hat{L}_B)$. The risk-averse buyer has an incentive to purchase cyber insurance if the expected cost under insurance is lower than the one without insurance:

$$(1 - r)\Phi(\hat{L}_B) + C_B(m) + C_I \leq \Phi(\hat{L}_B) + C_B(m).\tag{IR_B}$$

Note that we assume that the utility of the buyer does not include the penalty payment from the procurement contract and assume that the procurement does not involve an accountability contract. If so, we need to design the procurement contract and the insurance contract jointly as they are interdependent.

The mechanism design problem of the insurer is to determine the optimal premium rate C_I and the coverage r to maximize his profit. The insurer provides insurance only when the profit is non-negative. Thus, we have the following constraint.

$$J_I := C_I - r \cdot \hat{L}_B \geq 0\tag{IR_I}$$

We assume that the insurer is rational and risk-neutral so that they use the accurate value of the expected loss of the system when making decisions. The insurer solves the following optimization problem:

$$\begin{aligned}
\max_{r,\,C_I} \quad & J_I = C_I - r \cdot \hat{L}_B \\
\text{s.t.} \quad & (1 - r)\Phi(\hat{L}_B) + C_I \leq \Phi(\hat{L}_B) && (IR_B) \\
& C_I - r \cdot \hat{L}_B \geq 0 && (IR_I) \\
& r \in (0, 1] \\
& C_I \in \mathbb{R}^+
\end{aligned}\tag{4.48}$$

Combining the individual rationality constraints (IR_B) and (IR_I) with the biased utility function, we arrive at the following proposition.

Proposition 4.1 *The insurance contract is established between the insurer and the buyer if the premium $C_I \in \mathbb{R}^+$ and the coverage level $r \in (0, 1]$ satisfy*

$$\hat{L}_B \leq \frac{C_I}{r} \leq \Phi(\hat{L}_B) \tag{4.49}$$

This result shows that the ratio between the coverage level r and premium value C_I depends on the average loss of performance of the system and the risk aversion of the pursuer. Under this constraint, a risk-averse buyer will have the incentive to purchase the insurance. This provides a fundamental principle for designing the insurance policy.

4.6.3.3 Maximum Premium with Full Coverage

In this section, we discuss the maximum acceptable premium the risk-averse buyer is willing to pay. According to Proposition 4.1, the ratio between the coverage level and the premium C_I/r is bounded by the expected and biased loss of performance of the system. The maximum premium value can be achieved when the insurer is providing full coverage as $r = 1$.

Proposition 4.2 *The maximum acceptable premium for the buyer is achieved under the following insurance policy:*

$$r^* = 1, \qquad C_I^* = \Phi(\hat{L}_B). \tag{4.50}$$

Consider the PT risk aversion in (4.46). The maximum acceptable premium can be expressed as

$$C_I^* = \hat{\Phi}(\hat{L}_B) = (1 - P_A^m) \cdot \lambda(\Delta U_B)^\beta. \tag{4.51}$$

Proposition 4.3 *With full coverage $r = 1$, the maximum acceptable premium is higher than the unbiased expected loss when the performance difference is relatively small, as*

$$C_I^* \geq \hat{L}_B \qquad \text{if} \ \ 0 \leq \Delta U_B \leq \lambda^{\frac{1}{1-\beta}}.$$

Fig. 4.20 Maximum
acceptable premium under
different degrees of risk
aversion

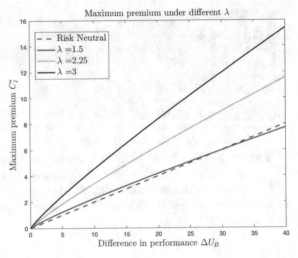

Fig. 4.21 Relationship
between accountability and
maximum acceptable
premium

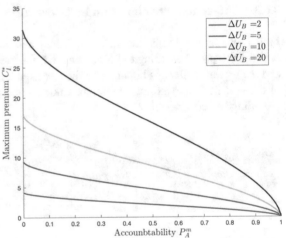

We first set $P_A^m = 0.8$, apply $\beta = 0.88$, $\zeta = 0.69$ in behavioral science literature and discuss the influence of loss aversion level λ on the maximum acceptable premium C_I^*, which is depicted in Fig. 4.20. The dotted line served as the baseline of the risk-neutral buyer, which represents the unbiased expected loss of performance. A larger value of λ indicates that the buyer is more risk-averse against the losses. The biased loss function is concave in ΔU_B because when the ΔU_B in performance is too high, a small increase in losses has little influence on the buyer's recognition.

Risk-averse buyers are sensitive to small losses, which provides the insurer an opportunity to take advantage of the risk aversion and charge for a higher premium. From the Fig. 4.21, the biased expected loss is greater than the unbiased one when ΔU_B is within the tolerable range for the buyer. This range coincides with the insurance purchase constraint in Proposition 4.1. If $\Delta U_B > \lambda^{\frac{1}{1-\beta}}$, we have $\Phi(\hat{L}_B) > \hat{L}_B$ and the buyer would not have the incentive to purchase cyber insurance anymore. This indicates that the insurer can increase the premium to its maximum acceptable value if the buyer participates in the insurance.

Proposition 4.4 *Cyber insurance is an incentive mechanism that encourages the buyer to have a more reliable accountability investigation.*

Another key result is that cyber insurance can increase the buyer's incentive to establish a more valid accountability investigation method. As described in (4.51), the maximum acceptable premium C_I^* has a negative correlation with respect to the accountability P_A^m. Let $\beta = 0.88$, $\lambda = 2.25$ and $\zeta = 0.69$ as the typical values in prospect theory, the influence of accountability investigation on the maximum acceptable premium is depicted in the following figure.

Figure 4.21 illustrates that a more reliable accountability investigation (larger P_A^m) can reduce the maximum premium of the insurance. The amount of reduction is higher if the performance differs more within two product types. If we consider the payoff function of the buyer under full insurance coverage. If the insurance company charges the maximum acceptable premium, we have

$$J_B(m, \delta) = C_B(m) + C_I^*. \tag{4.52}$$

The decrease in C_I will reduce the total payoff J_B of the buyer, resulting in a higher profit. In other words, cyber insurance provides incentives for the buyer to invest more in accountability investigation and establish a more reliable examination method to determine whether the supplier should be accountable for the incident.

4.6.3.4 Coverage Level with Given Premium

In this section, we discuss the coverage level r when the premium C_I is given. As demonstrated in Proposition 4.1, given a premium C_I, the insurance contract is established if

$$\frac{C_I}{\Phi(\hat{L}_B)} \le r \le \frac{C_I}{\hat{L}_B}. \tag{4.53}$$

This can be regarded as a constraint in the optimization problems for the buyer and the insurer.

Given C_I, the buyer's problem is to find the optimal coverage level that minimizes the total payoff under insurance.

$$\min_{r \in (0,1]} \quad J_B = (1 - r)\Phi(\hat{L}_B) + C_B(m) + C_I,$$

$$\text{s.t.} \quad \frac{C_I}{\Phi(\hat{L}_B)} \leq r \leq \frac{C_I}{\hat{L}_B}. \tag{OP_B}$$

Note that the buyer makes decision under biased expected loss, thus we use $\Phi(\hat{L}_B)$ in the objective function to represent her recognition. On the other hand, the insurer's problem is to find the optimal coverage level that maximizes his profit.

$$\max_{r \in (0,1]} \quad J_I = C_I - r\hat{L}_B,$$

$$\text{s.t.} \quad \frac{C_I}{\Phi(\hat{L}_B)} \leq r \leq \frac{C_I}{\hat{L}_B}. \tag{OP_I}$$

We assume that the insurer is rational and the expected loss in the objective function is unbiased.

By solving these two optimization problems (OP_B) and (OP_I), we find the optimal coverage levels for the buyer and the insurer as follows:

$$r_B^* = \max\left\{\frac{C_I}{\hat{L}_B}, 1\right\}, \qquad r_I^* = \min\left\{\frac{C_I}{\Phi(\hat{L}_B)}, 0\right\}.$$

The buyer prefers a larger coverage level achieved at the upper bound under the constraints, while the insurer favors a lower coverage level achieved at the lower bound. The result coincides with the fact that the insurance company and the buyer have a conflict of incentives in terms of the overall payoff. However, the individual preferences of both sides need to satisfy the constraint in (4.53) in order to establish the insurance contract in the first place.

Proposition 4.5 *Given the insurance premium C_I, the acceptable range of coverage level r will shift in the buyer's favor with more accountability P_A^m.*

Figure 4.22 illustrates the acceptable coverage level r when the performance difference $\Delta U_B = 6$ and given premium value $C_I = 2$. From the figure, both bounds of the coverage level increase with respect to the accountability P_A^m. This is because both $\hat{L}_B$ and $\Phi(\hat{L}_B)$ are decreasing functions in P_A^m. The phenomenon shows that a more reliable accountability investigation (larger P_A^m) benefits the buyer when he participates in cyber insurance. Since the insurance contract is only established under the constraint, the acceptable range of coverage level closer to 1 covers more portion of the losses in the system, thus reducing the payoff that the buyer needs to pay after a system malfunction.

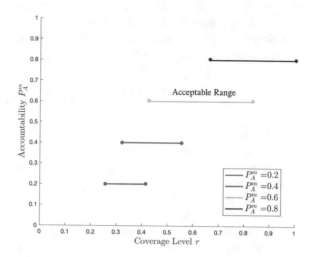

Fig. 4.22 Coverage level under different accountability ($\Delta U_B = 6$, $C_I = 2$)

4.6.3.5   Trade-Off Between Accountability Investment and Cyber Insurance

Lastly, we discuss the trade-off between the investment in accountability investigation and cyber insurance. From the previous discussion, a more reliable accountability investigation method (larger P_A^m) reduces the maximum acceptable premium C_I and increases the coverage level r. They result in a more favorable insurance plan for the buyer that mitigates the losses of performance due to the supplier. However, usually, the increase in P_A^m comes with a cost. It brings up the question: how much should we invest in accountability?

Suppose the cost to increase the accountability from P_A^m to $P_A^{m'}$ is C_n. This value represents the extra funding on accountability investigation. The total payoff of the buyer before (J_B) and after (J_B') accountability investment are

$$J_B = (1 - r)(1 - P_A^m)\Delta U_B + C_B(m) + C_I$$
$$J_B' = (1 - r')(1 - P_A^{m'})\Delta U_B + C_B(m) + C_I' + C_n$$
(4.54)

where r' and C_I' are the modified insurance plan. From previous discussions, we have arrived at $P_A^{m'} > P_A^m$, $r' > r$ and $C_I' < C_I$. The problem is to find the optimal investment such that

$$J_B' - J_B \leq 0.$$
(4.55)

The optimal investment depends on various factors such as the cost C_n, expected loss $\hat{L}_B$, the buyer's risk aversion, etc. We illustrate the trade-off between accountability investment and cyber insurance in the following example.

Example: Autonomous Truck Platooning

Consider the autonomous truck platooning example in Sect. 4.4.3. The accountability of the supplier takes the form

$$P_A^m(N) = 1 - Q\left(\frac{d}{2} + \frac{\ln(\tau)}{d}\right),\qquad(4.56)$$

where $d = N^{1/2}e_d/\sigma$. Normally, the sensor difference e_d, supplier's reputation ratio τ and observation variance σ^2 are already given. The only variable that is completely controlled by the investigator is the number of test N. From the analysis in the previous section, we know that $dP_A^m/dN \geq 0$. In order to reach a higher value of P_A^m, the buyer needs to increase the number of tests during the investigation, which is costly in general.

Consider the insurance plan with full coverage $r = 1$ and maximum premium C_I^* as described in Proposition 4.2. We assume the buyer obeys CARA risk aversion for the expected loss. Suppose the cost to conduct one test is c_n. The buyer would like to find out the optimal number of tests N that can minimize her payoff, which is

$$\min_N \quad J_B = (1 - r)\hat{L}_B + C_B(m) + C_I^* + N \cdot c_n$$
$$\qquad\qquad(4.57)$$
$$= C_B(m) + (1 - P_A^m(N))\phi(\Delta U_B) + N \cdot c_n$$

Figure 4.23 shows the optimal number of accountability tests with different test costs. When there is no cost to conduct one accountability test ($c_n = 0$), more tests are better for the buyer. Increasing the number of tests, in general, increases the accountability P_A^m. As $N \to \infty$, the accountability investigation can identify the untruthful supplier almost surely with $P_A^m \to 1$. In this case, the supplier will be penalized for the misinformation, and the payoff of the buyer will be

Fig. 4.23 Optimal number of test with different test cost

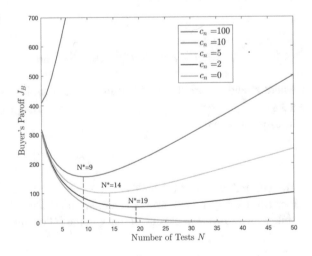

close to zero. When the cost of each test c_n increases, the optimal number of test N^* decreases. This illustrates the trade-off between accountability investigation and cyber insurance. Even though increasing the number of tests provides a more reliable test and reduce the insurance premium, the total investment would exceed the benefit after some point, causing unnecessary payoff for the buyer. Finally, if the investigation is too costly as $c_n = 100$, the buyer will never benefit from conducting an accountability investigation. It is better for the buyer to change to other comparatively low-cost investigation methods. By decreasing c_n, the buyer can find the optimal number of tests and achieve a lower payoff.

4.7 Conclusion

In this chapter, we have proposed a system-scientific framework to study the accountability in IoT supply chains and provided a holistic risk analysis technologically and socio-economically. We have developed stylized models and quantitative approaches to evaluate the accountability of the supplier. Two case studies have been used to demonstrate the model of accountability in the setting of autonomous truck platooning and ransomware in IoT supply chain.

We discuss the accountability investigation performance and design with a single supplier in the autonomous truck platooning case. From the parameter analysis, the reliability of the investigation can be improved with larger sensor error, more number of tests, and less observation variance. We have also showed the impact of the supplier's reputation on accountability investigation. A bad reputation will increase both accountability and wronged accountability during the investigation.

Using the smart lock case study, we have illustrated how to determine the accountability of the supplier in the IoT supply chain under a ransomware attack. A Neyman-Pearson test has been used to deal with suppliers with limited prior information. We have presented the model of the multi-stage accountability investigation with multiple suppliers in the supply chain and discussed the trade-off between detailed investigation and product replacement.

Contract design and cyber insurance are used as economic solutions to improve the cyber resilience in IoT supply chains. By designing contracts under incentive-compatibility and individual rationality constraints, the IoT end-user can penalize the accountable supplier and reduce his incentive of providing misinformation in the first place. Cyber insurance mitigates the loss of performance by transferring the risks to a third party. We have shown that cyber insurance is an incentive-compatible mechanism that facilitates a more reliable accountability investigation from the buyer side. However, the investigator needs to balance between the accountability investment and cyber insurance to achieve a higher payoff.

References

1. D.L. Farris, Target to pay nearly $40 million to settle with banks over data breach; total costs reach $290 million (2015). [Online]. Available: https://www.natlawreview.com/article/target-to-pay-nearly-40-million-to-settle-banks-over-data-breach-total-costs-reach
2. N. Manworren, J. Letwat, O. Daily, Why you should care about the target data breach. Bus. Horiz. **59**(3), 257–266 (2016)
3. T. Kieras, M.J. Farooq, Q. Zhu, Modeling and assessment of IoT supply chain security risks: the role of structural and parametric uncertainties, in *2020 IEEE Security and Privacy Workshops (SPW)* (IEEE, 2020), pp. 163–170
4. T. Kieras, M.J. Farooq, Q. Zhu, RIoTS: Risk analysis of IoT supply chain threats, in *2020 IEEE 6th World Forum on Internet of Things (WF-IoT)* (IEEE, 2020), pp. 1–6
5. T. Kieras, J. Farooq, Q. Zhu, I-SCRAM: A framework for IoT supply chain risk analysis and mitigation decisions. IEEE Access **9**, 29827–29840 (2021)
6. M.J. Farooq, Cyber-physical dynamic decision mechanisms for large scale Internet of things systems & networks, Ph.D. dissertation, New York University Tandon School of Engineering, 2020
7. L. Huang, Q. Zhu, Farsighted risk mitigation of lateral movement using dynamic cognitive honeypots, in *International Conference on Decision and Game Theory for Security* (Springer, 2020), pp. 125–146
8. J. Pawlick, E. Colbert, Q. Zhu, A game-theoretic taxonomy and survey of defensive deception for cybersecurity and privacy. ACM Comput. Surv. (CSUR) **52**(4), 82 (2019)
9. J. Pawlick, Q. Zhu, *Game Theory for Cyber Deception: From Theory to Applications* (Springer Nature, 2021)
10. L. Huang, Q. Zhu, Duplicity games for deception design with an application to insider threat mitigation. IEEE Trans. Inf. Forens. Secur. **16**, 4843–4856 (2021)
11. Q. Zhu, T. Başar, Game-theoretic methods for robustness, security, and resilience of cyber-physical control systems: games-in-games principle for optimal cross-layer resilient control systems. Control Syst. IEEE **35**(1), 46–65 (2015)
12. L. Huang, Q. Zhu, A dynamic games approach to proactive defense strategies against advanced persistent threats in cyber-physical systems. Comput. Secur. **89**, 101660 (2020)
13. Q. Zhu, Z. Xu, *Cross-layer Design for Secure and Resilient Cyber-physical Systems* (Springer, 2020)
14. Y. Huang, L. Huang, Q. Zhu, Reinforcement learning for feedback-enabled cyber resilience. Preprint. arXiv:2107.00783 (2021)
15. C.A. Kamhoua, C.D. Kiekintveld, F. Fang, Q. Zhu, *Game Theory and Machine Learning for Cyber Security* (Wiley, 2021)
16. L. Huang, Q. Zhu, Adaptive honeypot engagement through reinforcement learning of semi-Markov decision processes, in *International Conference on Decision and Game Theory for Security* (Springer, 2019), pp. 196–216
17. R. Zhang, Q. Zhu, Y. Hayel, A bi-level game approach to attack-aware cyber insurance of computer networks. IEEE J. Sel. Areas Commun. **35**(3), 779–794 (2017)
18. C.J. Fung, Q. Zhu, FACID: A trust-based collaborative decision framework for intrusion detection networks. Ad Hoc Netw. **53**, 17–31 (2016). [Online]. Available: http://www.sciencedirect.com/science/article/pii/S1570870516302062
19. M.H. Manshaei, Q. Zhu, T. Alpcan, T. Bacşar, J.P. Hubaux, Game theory meets network security and privacy. ACM Comput. Surv. (CSUR) **45**(3), 25 (2013)
20. Q. Zhu, C. Fung, R. Boutaba, T. Başar, GUIDEX: A game-theoretic incentive-based mechanism for intrusion detection networks. IEEE J. Sel. Areas Commun. **30**(11), 2220–2230 (2012)
21. Q. Zhu, H. Tembine, T. Başar, Network security configurations: A nonzero-sum stochastic game approach, in *Proceedings of the 2010 American Control Conference* (IEEE, 2010), pp. 1059–1064

22. T. Zhang, Q. Zhu, Strategic defense against deceptive civilian GPS spoofing of unmanned aerial vehicles, in *International Conference on Decision and Game Theory for Security* (Springer, 2017), pp. 213–233
23. Q. Zhu, Z. Yuan, J.B. Song, Z. Han, T. Başar, Interference aware routing game for cognitive radio multi-hop networks. IEEE J. Sel. Areas Commun. **30**(10), 2006–2015 (2012)
24. Q. Zhu, J.B. Song, T. Başar, Dynamic secure routing game in distributed cognitive radio networks, in *Global Telecommunications Conference (GLOBECOM 2011), 2011 IEEE* (IEEE, 2011), pp. 1–6
25. Q. Zhu, H. Li, Z. Han, T. Başar, A stochastic game model for jamming in multi-channel cognitive radio systems, in *ICC* (2010), pp. 1–6
26. Q. Zhu, W. Saad, Z. Han, H.V. Poor, T. Başar, Eavesdropping and jamming in next-generation wireless networks: A game-theoretic approach, in *Military Communications Conference (MILCOM), 2011* (IEEE, 2011), pp. 119–124
27. Q. Zhu, Z. Yuan, J.B. Song, Z. Han, T. Başar, Dynamic interference minimization routing game for on-demand cognitive pilot channel, in *Global Telecommunications Conference (GLOBECOM 2010), 2010 IEEE* (IEEE, 2010), pp. 1–6
28. J. Pawlick, E. Colbert, Q. Zhu, Modeling and analysis of leaky deception using signaling games with evidence. IEEE Trans. Inf. Forens. Secur. **14**(7), 1871–1886 (2018)
29. J. Zheng, D.A. Castañón, Dynamic network interdiction games with imperfect information and deception, in *2012 IEEE 51st IEEE Conference on Decision and Control (CDC)* (IEEE, 2012), pp. 7758–7763
30. Q. Zhu, A. Clark, R. Poovendran, T. Başar, Deceptive routing games, in *2012 IEEE 51st IEEE Conference on Decision and Control (CDC)* (IEEE, 2012), pp. 2704–2711
31. K. Horák, Q. Zhu, B. Bošanský, Manipulating adversary's belief: A dynamic game approach to deception by design for proactive network security, in *International Conference on Decision and Game Theory for Security* (Springer, 2017), pp. 273–294
32. L. Huang, Q. Zhu, A dynamic games approach to proactive defense strategies against advanced persistent threats in cyber-physical systems. CoRR, vol. abs/1906.09687 (2019). [Online]. Available: http://arxiv.org/abs/1906.09687
33. Q. Zhu, S. Rass, On multi-phase and multi-stage game-theoretic modeling of advanced persistent threats. IEEE Access **6**, 13958–13971 (2018)
34. J. Chen, C. Touati, Q. Zhu, A dynamic game analysis and design of infrastructure network protection and recovery. ACM SIGMETRICS Perform. Eval. Rev. **45**(2), 128 (2017)
35. J. Chen, Q. Zhu, Interdependent strategic cyber defense and robust switching control design for wind energy systems, in *Power & Energy Society General Meeting, 2017 IEEE* (IEEE, 2017), pp. 1–5
36. S. Rass, S. Schauer, S. König, Q. Zhu, *Cyber-Security in Critical Infrastructures: A Game-Theoretic Approach*. Advanced Sciences and Technologies for Security Applications (Springer, 2020)
37. C. Rieger, I. Ray, Q. Zhu, M. Haney, *Industrial Control Systems Security and Resiliency: Practice and Theory*. Advances in Information Security (Springer, 2019)
38. Q. Zhu, T. Başar, Robust and resilient control design for cyber-physical systems with an application to power systems, in *2011 50th IEEE Conference on Decision and Control and European Control Conference* (IEEE, 2011), pp. 4066–4071
39. Q. Zhu, L. Bushnell, T. Başar, Resilient distributed control of multi-agent cyber-physical systems, in *Control of Cyber-Physical Systems* (Springer, 2013), pp. 301–316
40. F. Miao, Q. Zhu, M. Pajic, G.J. Pappas, A hybrid stochastic game for secure control of cyber-physical systems. Automatica **93**, 55–63 (2018)
41. Z. Xu, Q. Zhu, A cyber-physical game framework for secure and resilient multi-agent autonomous systems, in *2015 IEEE 54th Annual Conference on Decision and Control (CDC)* (IEEE, 2015), pp. 5156–5161
42. J. Chen, C. Touati, Q. Zhu, Optimal secure two-layer IoT network design. IEEE Trans. Control Netw. Syst. **7**(1), 398–409 (2019)

43. Q.D. La, T.Q. Quek, J. Lee, A game theoretic model for enabling honeypots in IoT networks, in *2016 IEEE International Conference on Communications (ICC)* (IEEE, 2016), pp. 1–6
44. J. Chen, Q. Zhu, Interdependent strategic security risk management with bounded rationality in the Internet of things. IEEE Trans. Inf. Forens. Secur. **14**(11), 2958–2971 (2019)
45. J. Chen, C. Touati, Q. Zhu, A dynamic game approach to designing secure interdependent IoT-enabled infrastructure network. IEEE Trans. Netw. Sci. Eng. **8**(3), 2601–2612 (2021)
46. J. Chen, Q. Zhu, *A Game-and Decision-Theoretic Approach to Resilient Interdependent Network Analysis and Design* (Springer, 2019)
47. T. Börgers, D. Krahmer, *An Introduction to the Theory of Mechanism Design* (Oxford University Press, USA, 2015)
48. R.B. Myerson, Perspectives on mechanism design in economic theory. Am. Econ. Rev. **98**(3), 586–603 (2008)
49. H. Nissenbaum, Computing and accountability. Commun. ACM **37**(1), 72–81 (1994)
50. J. Feigenbaum, A.D. Jaggard, R.N. Wright et al., *Accountability in Computing: Concepts and Mechanisms* (Now Publishers, 2020)
51. J. Feigenbaum, A.D. Jaggard, R.N. Wright, Open vs. closed systems for accountability, in *Proceedings of the 2014 Symposium and Bootcamp on the Science of Security* (2014), pp. 1–11
52. R. Künnemann, I. Esiyok, M. Backes, Automated verification of accountability in security protocols, in *2019 IEEE 32nd Computer Security Foundations Symposium (CSF)* (IEEE, 2019), pp. 397–39716
53. J. Zou, Y. Wang, K.J. Lin, A formal service contract model for accountable SAAS and cloud services, in *2010 IEEE International Conference on Services Computing* (IEEE, 2010), pp. 73–80
54. R. Avenhaus, B. Von Stengel, and S. Zamir, Inspection games, *Handbook of game theory with economic applications* 3, pp. 1947–1987, 2002.
55. T. Zhang, Q. Zhu, Hypothesis testing game for cyber deception, in *International Conference on Decision and Game Theory for Security* (Springer, 2018), pp. 540–555
56. G. Peng, Q. Zhu, Sequential hypothesis testing game, in *2020 54th Annual Conference on Information Sciences and Systems (CISS)* (IEEE, 2020), pp. 1–6
57. J. Blocki, N. Christin, A. Datta, A.D. Procaccia, A. Sinha, Audit games, in *Twenty-Third International Joint Conference on Artificial Intelligence* (2013)
58. S. Rass, S. Schauer, S. König, Q. Zhu, Optimal inspection plans, in *Cyber-Security in Critical Infrastructures* (Springer, 2020), pp. 179–209
59. S. Rass, Q. Zhu, GADAPT: a sequential game-theoretic framework for designing defense-in-depth strategies against advanced persistent threats, in *International Conference on Decision and Game Theory for Security* (Springer, 2016), pp. 314–326
60. R.B. Myerson, Optimal auction design. Math. Oper. Res. **6**(1), 58–73 (1981)
61. M.J. Farooq, Q. Zhu, Optimal dynamic contract for spectrum reservation in mission-critical UNB-IoT systems, in *2018 16th International Symposium on Modeling and Optimization in Mobile, Ad Hoc, and Wireless Networks (WiOpt)* (IEEE, 2018), pp. 1–6
62. T. Zhang, Q. Zhu, Optimal two-sided market mechanism design for large-scale data sharing and trading in massive IoT networks. Preprint. arXiv:1912.06229 (2019)
63. T. Zhang, Q. Zhu, On incentive compatibility in dynamic mechanism design with exit option in a Markovian environment. Dyn. Games Appl. **12**, 701–745 (2022)
64. J. Chen, Q. Zhu, Security as a service for cloud-enabled Internet of controlled things under advanced persistent threats: a contract design approach. IEEE Trans. Inf. Forens. Secur. **12**(11), 2736–2750 (2017)
65. R. Zhang, Q. Zhu, FlipIn:a game-theoretic cyber insurance framework for incentive-compatible cyber risk management of internet of things. IEEE Trans. Inf. Forens. Secur. **15**, 2026–2041 (2019)
66. L. Huang, Q. Zhu, Dynamic bayesian games for adversarial and defensive cyber deception, in *Autonomous Cyber Deception* (Springer, 2019), pp. 75–97

67. S. Jajodia, A.K. Ghosh, V. Swarup, C. Wang, X.S. Wang, *Moving Target Defense: Creating Asymmetric Uncertainty for Cyber Threats*, vol. 54 (Springer Science & Business Media, 2011)

68. Q. Zhu, T. Başar, Game-theoretic approach to feedback-driven multi-stage moving target defense, in *International Conference on Decision and Game Theory for Security* (Springer, 2013), pp. 246–263

69. Z. Qian, J. Fu, Q. Zhu, A receding-horizon MDP approach for performance evaluation of moving target defense in networks, in *2020 IEEE Conference on Control Technology and Applications (CCTA)* (IEEE, 2020), pp. 1–7

70. B.C. Levy, Binary and Mary hypothesis testing, in *Principles of Signal Detection and Parameter Estimation* (Springer, 2008), pp. 1–57

71. T.D. Wickens, *Elementary Signal Detection Theory* (Oxford University Press, 2001)

72. J.H. Shapiro, Bounds on the area under the ROC curve. JOSA A **16**(1), 53–57 (1999)

73. J.N. Tsitsiklis, Decentralized detection, in *Advances in Statistical Signal Processing, Signal Detection*, ed. by Poor, Thomas, vol. 2, (JAI Press, 1990)

74. K. C. Nguyen, T. Alpcan, and T. Basar, Distributed hypothesis testing with a fusion center: The conditionally dependent case, in *2008 47th IEEE Conference on Decision and Control* (IEEE, 2008), pp. 4164–4169

75. W.H. Organization et al., Global status report on road safety 2018: summary, World Health Organization, Tech. Rep. (2018)

76. C. Stöckle, W. Utschick, S. Herrmann, T. Dirndorfer, Robust design of an automatic emergency braking system considering sensor measurement errors, in *2018 21st International Conference on Intelligent Transportation Systems (ITSC)* (IEEE, 2018)

77. M. Wang, W. Daamen, S.P. Hoogendoorn, B. van Arem, Rolling horizon control framework for driver assistance systems. part I: Mathematical formulation and non-cooperative systems. Transp. Res. Part C Emerg. Technol. **40**, 271–289 (2014)

78. D. Braue, Global ransomware damage costs predicted to exceed $265 billion by 2031, 2021, accessed: July 20, 2021. [Online]. Available: https://cybersecurityventures.com/global-ransomware-damage-costs-predicted-to-reach-250-billion-usd-by-2031/

79. D. Geneiatakis, I. Kounelis, R. Neisse, I. Nai-Fovino, G. Steri, G. Baldini, Security and privacy issues for an IoT based smart home, in *2017 40th International Convention on Information and Communication Technology, Electronics and Microelectronics (MIPRO)* (IEEE, 2017), pp. 1292–1297

80. E. Cartwright, J. Hernandez Castro, A. Cartwright, To pay or not: game theoretic models of ransomware. J. Cybersecur. **5**(1), tyz009 (2019)

81. J. Neyman, E.S. Pearson, IX. On the problem of the most efficient tests of statistical hypotheses. Philos. Trans. R. Soc. Lond. A **231**(694–706), 289–337 (1933). Containing Papers of a Mathematical or Physical Character

82. J. Boyens, C. Paulsen, R. Moorthy, N. Bartol, Supply chain risk management practices for federal information systems and organizations (2015). [Online]. Available: https://nvlpubs.nist.gov/nistpubs/SpecialPublications/NIST.SP.800-161.pdf

83. D. Kovaleski, Bill that requires security standards for government purchases of iot devices signed into law (2020). [Online]. Available: https://homelandprepnews.com/stories/58555-bill-that-requires-security-standards-for-government-purchases-of-iot-devices-signed-into-law/

84. R.L. Kelly, Text - h.r.1668 - 116th congress (2019-2020): Internet of things cybersecurity improvement act of 2020 (2020). [Online]. Available: https://www.congress.gov/bill/116th-congress/house-bill/1668/text

85. Executive order on improving the nation's cybersecurity, May (2021). [Online]. Available: https://www.whitehouse.gov/briefing-room/presidential-actions/2021/05/12/executive-order-on-improving-the-nations-cybersecurity/

86. J. Hecht, Lidar for self-driving cars. Opt. Photonics News **29**(1), 26–33 (2018)

87. N.A. of Insurance Commissioners (NAIC), Report on the cybersecurity insurance market (2021). Accessed 20 Oct 2021. [Online]. Available: https://content.naic.org/sites/default/files/index-cmte-c-Cyber_Supplement_2020_Report.pdf

88. B. Cashell, W.D. Jackson, M. Jickling, B. Webel, The economic impact of cyber-attacks, Congressional research service documents, CRS RL32331 (Washington DC), 2 (2004)
89. R.P. Majuca, W. Yurcik, J.P. Kesan, The evolution of cyberinsurance. Preprint. cs/0601020 (2006)
90. A. Marotta, F. Martinelli, S. Nanni, A. Orlando, A. Yautsiukhin, Cyber-insurance survey. Comput. Sci. Rev. **24**, 35–61 (2017)
91. R. Zhang, Q. Zhu, Optimal cyber-insurance contract design for dynamic risk management and mitigation. IEEE Trans. Comput. Soc. Syst. (2021)
92. R. Zhang, Strategic cyber data risk management over networks: from proactive defense to cyber insurance, Ph.D. dissertation, New York University Tandon School of Engineering, 2020
93. R. Pal, L. Golubchik, K. Psounis, P. Hui, Will cyber-insurance improve network security? a market analysis, in *IEEE INFOCOM 2014-IEEE Conference on Computer Communications* (IEEE, 2014), pp. 235–243
94. R. Böhme, G. Schwartz et al., Modeling cyber-insurance: Towards a unifying framework, in *WEIS* (2010)
95. P. Radanliev, D. De Roure, S. Cannady, R. Mantilla Montalvo, R. Nicolescu, M. Huth, Analysing IoT cyber risk for estimating IoT cyber insurance, in *Living in the Internet of Things: Cybersecurity of the IoT-2018. IET Conference Proceedings* (The Institution of Engineering and Technology, London, 2018), pp. 1–9
96. D. Kahneman, A. Tversky, Prospect theory: An analysis of decision under risk, in *Handbook of the Fundamentals of Financial Decision Making: Part I* (World Scientific, 2013), pp. 99–127

Chapter 5
Computational Tools

Abstract Computational tools are critical in measuring systemic risk and assisting with mitigation decisions. While several software packages and tools are available for the traditional supply chain management solutions, there is no dedicated tool to assist with the cyber supply chain risk analysis and mitigation. This chapter introduces a software tool named I-SCRAM that is specifically designed to allow the vendor risk assessment of IT, OT, and IoT systems and assist with cost-effective risk minimizing selection of vendors. Case studies are provided to help illustrate the utility of the developed tool.

5.1 Introduction to I-SCRAM: A Software Tool for IoT SCRM

This chapter introduces I-SCRAM, a software tool for conducting supply chain risk analysis and mitigation decisions. The tool itself aims to implement the framework that has already been described, and so after a short introduction, this chapter will focus on a description of the interface of the tool illustrated by two case studies.

5.1.1 Supply Chain Risk Analysis and Mitigation

As described earlier, the framework utilizes a graph-based system model, where nodes are suppliers and components, and edges represent dependencies of relevance for the system's functions and security. An additional dimension of modeling consists in assigning nodes a particular logic function that determines the state of the node based on the states of its dependencies. With these abstractions at hand, the I-SCRAM framework can be employed to model the risks in a great variety of systems.

In addition to the structural aspects of the system model, I-SCRAM presupposes that probabilities can be assigned to each node that represents the best available assessment of the individual node's likelihood to introduce risk. These risks may

© The Author(s), under exclusive license to Springer Nature Switzerland AG 2022
T. Kieras et al., *IoT Supply Chain Security Risk Analysis and Mitigation*,
SpringerBriefs in Computer Science, https://doi.org/10.1007/978-3-031-08480-5_5

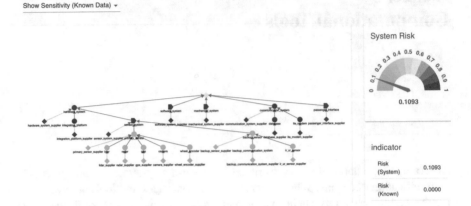

Select Metric
Show Sensitivity (Known Data) ▾

Fig. 5.1 The ISCRAM analysis page featuring case study 1: autonomous vehicle

be risks of the individual component compromise due to, for example, poorly implemented cryptography, or they may be risks of non-security-related failures in functionality. While this offers a great deal of flexibility, the requirement for a coherent analysis is that the occurrence of whatever event that is modeled must be a contributor to the failures of adjacent nodes in the system.

Because a central problem in carrying out this analysis in practice relates to the availability and reliability of quantitative risk assessments of individual components and suppliers, an important feature of I-SCRAM's analysis consists in allowing the discovery of systemic sources of risk independent of the particular probabilities at each node. We will see in the discussions that follow that the interface of the I-SCRAM tool, shown in Fig. 5.1, allows for the user to carry out some analysis even when the probability values are uncertain. With these general observations in mind, what follows will be a description of the chief components of the I-SCRAM tool and its capabilities.

5.1.1.1   I-SCRAM Software Components

The chief components of the I-SCRAM tool are a user interface available in a web browser, a server that conducts the analysis itself, and a data model that is provided by the user as input. The I-SCRAM server is written in Python, with source code available at (https://github.com/tkieras/iscram). Future developments of I-SCRAM can be obtained at [https://www.i-scram.com/]

5.1.1.2   User Interface

The user interface is implemented to allow the chief capabilities of the tool to be accessible from a web browser. The interface's main features are data visualizations to show the system graph and the various metrics that are computed by the server. Separate views are available for risk analysis and mitigation decisions. Editing the system graph is also possible.

5.1.1.3 Data Model

The data model captures the problem data that the user must provide as input. All operations of the I-SCRAM tool require this data as input. The data is provided in a JSON object which is loaded into the tool from a file. The data can also be edited in the interface. The input has three main parts: 'system_graph', 'data', and 'preferences'. Details of the schema required in these fields can be found by running the server and viewing the OpenAPI documentation, but a general overview here will be helpful in understanding the nature of the input required. In general, the 'system_graph' portion of the data specifies the structural aspects of the system itself, essentially nodes and edges. The 'data' portion then adds specific probabilities and attributes to nodes or edges. These two portions are separated to allow more efficient operations when the structure of the system is fixed and what varies more frequently are the probabilities and attributes specified in the data portion.

5.1.1.4 Server

The server itself implements an HTTP API which returns one of several different metrics based on the 'system_graph' and 'data' input. While the primary use case consists in using the I-SCRAM tool via a browser-based client, it is possible to make requests to the server using any HTTP client. Likewise, the server's core operations can be invoked via a CLI interface independent from the client-server architecture.

The major functions provided by the I-SCRAM server are:

- the Birnbaum importances of nodes based on the provided risk values,
- the Birnbaum structural importances of nodes based on unknown risk values,
- the Birnbaum importances of attributes,
- the total system risk,
- the Birnbaum importances of supplier attributes,
- the recommended supplier choices given a specified budget.

When displaying options to the user, a distinction is presented between sensitivity based on known risk values and sensitivity based on unknown risk values. In the first case, what is computed is the Birnbaum importance measure, and in the second case it is the Birnbaum structural importance measure as described earlier in this book. The distinction is particularly useful for the I-SCRAM tool. In the first case, the importances of nodes are largely determined by the risk values provided in the data input. These risk values represent the output of a process of estimation of risks performed by analysts and based on observed patterns of failure. These risk values may or may not be highly reliable, and in many cases will be a best-effort score. With this in mind, the Birnbaum structural importance can provide a needed alternative perspective. In this measure, the risk values for each node are assumed to be unknown, i.e., the nodes are equally likely to fail or not fail. Given these values, the computed importances reflect only the structure of the system graph. Depending on the use case and the individual availability of strong risk values for each node, either of the two importances would be more influential when assessing optimal risk-mitigating decisions.

To assist in clarifying the nodes of higher importances, an optional normalization may be applied to these importances. In this case what is sought by the analysis is chiefly the relative ranking of nodes by importance, with a view to making risk mitigating decisions.

5.1.1.5   Implementation of Core Operations

Several technical challenges were presented in developing the I-SCRAM tool and it will be helpful here to describe briefly the approach taken to implement the core operations of the tool efficiently. The first challenge was to compute the Birnbaum measures on larger system graphs. As described earlier in this book, a simple algorithm for this computation utilizes the MOCUS algorithm for finding the minimal cutsets in a fault tree [1]. After these cutsets are found the calculation can be computed efficiently. However, there may be a very large number of cutsets and this leads to computational challenges in terms of memory usage and running time. The alternate approach explored in recent research is the use of the Binary Decision Diagram (BDD) data structure [2], which has been used to evaluate fault trees of significant size [3–6]. A BDD-based approach is used in I-SCRAM to support larger system graphs and minimize the computational resources required. When a user first loads a system graph into the tool, a BDD is constructed using the CUDD software library (https://web.archive.org/web/20150215010018/http://vlsi.colorado.edu/~fabio/CUDD/cuddIntro.html). The data structure is cached on the server and used whenever there is a need to evaluate the system risk or compute sensitivity metrics for the system graph in question.

The second aspect of I-SCRAM's implementation that should be described briefly is the handling of optimization when solving the supplier choice problem. The tools used for solving the problem as formulated earlier in this book are the *pyomo* Python package for interfacing with specialized optimization tools [7, 8], and the *Couenne* solver developed by the COIN-OR project (http://www.coin-or.org/Couenne).

5.2   Case Study 1: Autonomous Vehicle

In this case study, the system in question is an Autonomous Vehicle with a structure described by diagram in Fig. 5.2, which is inspired from the study conducted in [9].

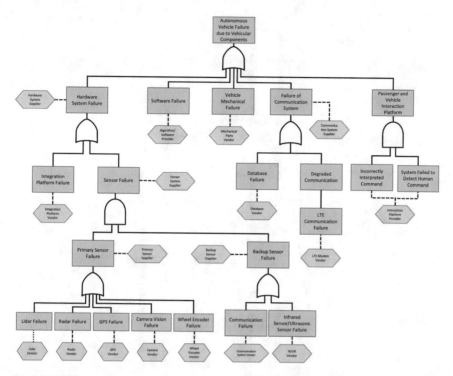

Fig. 5.2 Attack Tree model for an autonomous vehicle failure due to vehicular components

After translating this diagram into a system graph following the requirements described earlier, the data can be loaded in to the I-SCRAM tool for analysis. The analysis interface is shown in Fig. 5.3. The main component of the interface is a graph visualization where nodes are colored according to one of several methods. Selected in this figure is a coloring by the individual risk values. In Table 5.1 a subset of the risk data is presented.

In Fig. 5.4 we see an alternate coloring method, by the Birnbaum importance measure (or sensitivity using known data). Data for this visualization is presented in Table 5.2. As we develop these case studies more will be said on the particulars of the case, but it can be seen clearly that certain nodes are immediately highlighted as being of higher importance than others.

It is instructive to contrast these two figures because when considering nodes individually they roughly appear to be without significant difference in their risks. However, the relative importances of the nodes given their location in the system structure vary widely, and this variance would not be captured simply by considering nodes individually.

Given the data presented, the interface will present the top five suppliers and components according to whatever metric is selected. In Fig. 5.5 the Birnbaum importance is once again selected. In this case study the nodes roughly fall into one

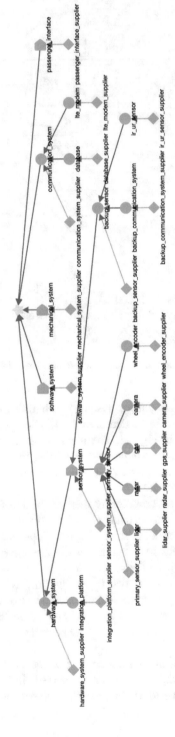

Fig. 5.3 System graph for the Autonomous Vehicle system visualized with green colors indicating low risks on each component and supplier

Table 5.1 Risk values for case study 1: Autonomous Vehicle

mechanical_system	0.001
communication_system	0.0100
passenger_interface	0.0150
lte_modem	0.0020
gps_supplier_alt	0.0030
gps	0.0020
indicator	0.0000
sensor_system	0.0030
camera_supplier_alt	0.0013
database_supplier_alt	0.0100
camera	0.0100
ir_ur_sensor	0.0010
backup_sensor_supplier_alt	0.0093
wheel_encoder	0.0010
integration_platform	0.0150
radar_supplier	0.0009
passenger_interface_supplier	0.0012
ir_ur_sensor_supplier_alt	0.0012
communication_system_supplier_alt	0.0015
mechanical_system_supplier_alt	0.0013
radar_supplier_alt	0.0012
backup_communication_system_supplier_alt	0.0013
mechanical_system_supplier	0.0019
wheel_encoder_supplier	0.0150
backup_sensor_supplier	0.0015
sensor_system_supplier	0.0100
integration_platform_supplier_alt	0.0013
communication_system_supplier	0.0015
primary_sensor	0.0010
lte_modem_supplier	0.0015
primary_sensor_supplier_alt	0.0010
lte_modem_supplier_alt	0.0020
wheel_encoder_supplier_alt	0.0008

of two categories, and they are either highly important or of minimal importance based on their position in this relatively uniform system graph. As we will see in the following case study, this is not always the case.

I-SCRAM also features a helpful analysis of suppliers based on attributes that are assigned to them. These attributes must be binary valued and are intended to capture a wide variety of features that may be of interest to the analyst. In this example, each supplier is tagged with the attributes: certified, small, domestic, and existing_contract. If the analyst can provide attributes such as these for each supplier, the attributes can be analyzed across the system in question.

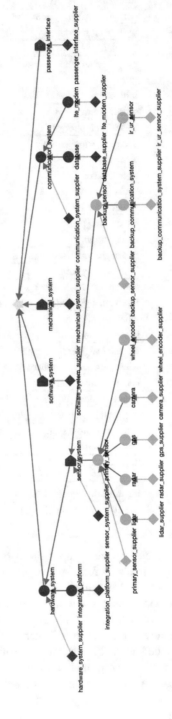

Fig. 5.4 System graph for the Autonomous Vehicle system visualized with colors indicating varying importance of each component and supplier

Table 5.2 Birnbaum importances for case study 1: Autonomous Vehicle

hardware_system	0.990
hardware_system_supplier	0.983
software_system	1.000
software_system_supplier	0.982
mechanical_system	0.981
mechanical_system_supplier	0.982
communication_system	0.990
communication_system_supplier	0.981
integration_platform	0.995
integration_platform_supplier	0.981
sensor_system	0.983
sensor_system_supplier	0.990
primary_sensor	0.019
primary_sensor_supplier	0.019
lidar	0.020
lidar_supplier	0.019
radar	0.019
radar_supplier	0.019
gps	0.019
gps_supplier	0.019
camera	0.019
camera_supplier	0.019
wheel_encoder	0.019
wheel_encoder_supplier	0.019
passenger_interface	0.995
passenger_interface_supplier	0.981
database	0.995
database_supplier	0.980
lte_modem	0.982
lte_modem_supplier	0.981
backup_sensor	0.068
backup_sensor_supplier	0.067
backup_communication_system	0.067
backup_communication_system_supplier	0.067
ir_ur_sensor	0.067
ir_ur_sensor_supplier	0.067

In Fig. 5.6, we see a simple visualization of each attribute according to its representation in the system. In other words, what fraction of suppliers have each value for the attribute. We see for example that many more suppliers have the existing_contract attribute than not. Clearly for the attribute to have a stable meaning it must correspond to some policy definition that can be resolved into a binary attribute given observed data about the supplier.

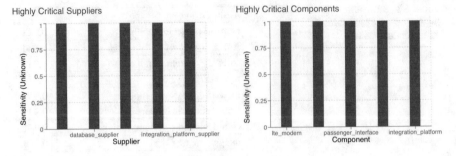

Fig. 5.5 Top five components and suppliers according to their impact on system risk

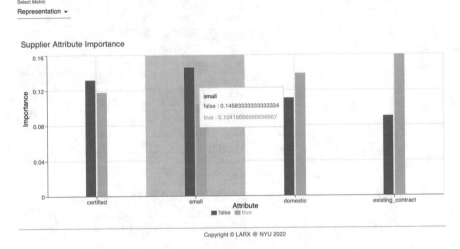

Fig. 5.6 Supplier attributes, fractions of each value represented

In Fig. 5.7, we see an alternate perspective on the attributes, using the Birnbaum importance measure described earlier. The importance of the attribute is computed by treating the entire set of suppliers with an attribute as being a unit and calculating the relative system risk when this unit is varied in its functionality. We see in this case study that the attributes all have similar importance, and that the values of each attribute are similar to each other. This is indicative that these attributes in this system are well distributed across suppliers in varying positions within the system. No attribute simply captures a larger share of the system risk than any other.

5.3 Case Study 2: Industrial Control System

In this case study, we will consider the example of an Industrial Control System with a structure described by diagram show in Fig. 5.8. The characteristics of this system will be different from the Autonomous Vehicle as will be seen in the more

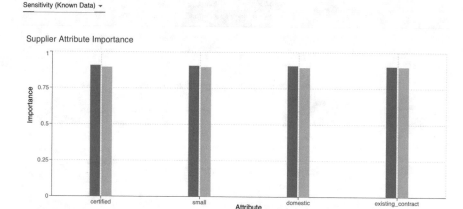

Select Metric
Sensitivity (Known Data) ▾

Fig. 5.7   System risk sensitivity for each supplier attribute value

varied node importances. We will also examine the decision support features of the I-SCRAM tool by considering how best to minimize risk given a certain budget and given various choices for each supplier.

In Fig. 5.9 we see the visualization of the Birnbaum importances for suppliers and components in this case study. The values computed here, again, take into account the risk values provided as user input. The data for this visualization is found in Table 5.3. The top five components and suppliers according to their importance are found in Fig. 5.10. We see here that there is a clear indication, given the data and analysis, that nodes that are contributors to the Firewall and Corporate Network are the most critical nodes, as well as the Sup_PLC_top node. The other nodes appear to be relatively insignificant.

It is useful to contrast the above analysis with the more cautious Birnbaum structural importance measure, which again does not take into account the user-provided risk values. This visualization is provided in Fig. 5.11 and the data in Table 5.4. The top five suppliers and components are shown in Fig. 5.12. In this case, we see a similar pattern, though the contrast between the two sets of nodes is less significant. Given that this measure is computed without reliance on the 'best effort' risk values, there is certain robustness to the analysis that might be lacking in comparison to the earlier results. Because these risk values are typically subject to various unknowns, the alignment of these two measures can lend weight to the decision to allocate resources on the Firewall and Corporate Network as well as their suppliers. Yet this measure is also more cautious and would suggest that the other nodes in the system, highlighted here in yellow though pictured as green in the earlier figure, deserve attention as well (Figs. 5.13 and 5.14).

As described above, I-SCRAM allows suppliers to be assigned attributes. In this case, we see that the attributes in question (as illustrated in Fig. 5.13) are more unequally distributed through the system than in the prior case study. The

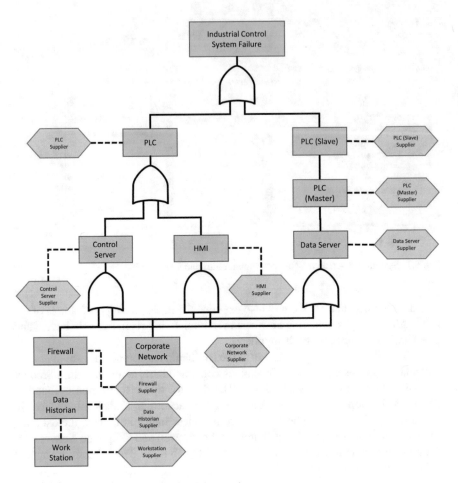

Fig. 5.8 Attack tree model for an industrial control system

importances shown in Fig. 5.14, however, appear quite similar and without much variation again between the values of each attribute. The contrast between the two graphs shows that although many more suppliers are marked as 'domestic' than otherwise, the system is not considerably more sensitive to the 'domestic' attribute than otherwise. A recurring theme in this analysis is that the structure of the system in question plays a significant role in determining the importance of each node, much more than simpler metrics such as representation would suggest.

In this case study, we will also examine the decision support features of the I-SCRAM tool, pictured in Fig. 5.15. The graph pictured here appears more complex because each node has several possible suppliers pictured. The importances and risks can be visualized as usual, though it is important to mention that only one supplier of each component is active at any given time. In other words, suppliers that are merely potential suppliers of a component are pictured with an edge to

Select Metric

Show Sensitivity (Known Data) ▾

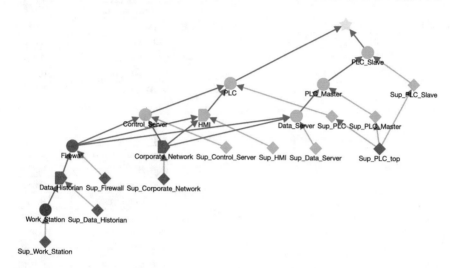

Fig. 5.9 Components and suppliers colored according to Birnbaum importances using the scheme shown in Fig. 5.1

the component, but they do not yet contribute any risk to the component. The optimization interface allows the user to specify a budget and a degree of supply chain risk tolerance . These correspond to budget and α values described in earlier chapters. Data for the supplier choice problem pictured here is provided in Table 5.5. Essentially for each component, there are various suppliers, where a higher price is correlated with a lower risk.

The result of the optimization is shown by the updated risk and importance values in the graph, as well as the updated chart of the highly critical suppliers and components. What we see in Fig. 5.16 are the top five components after a low budget optimization. Given few resources, we end up choosing the riskier suppliers. In Fig. 5.17 the result is shown after a higher budget optimization, featuring a lower importance measure for many of the suppliers and components. These contrasting plots show that the higher budget was able to not only reduce the system risk but also reduce the dependence of the system on particular, highly critical components and suppliers.

5.4   Conclusions and Outlooks

In this chapter, we have presented an introduction and usage of a software tool named I-SCRAM, designed to provide decision support for risk mitigation from

Table 5.3 Birnbaum
importances for Case Study
2: Industrial Control System

PLC	0.240
PLC_Slave	0.231
HMI	0.243
Control_Server	0.248
Firewall	0.944
PLC_Master	0.236
Data_Server	0.241
Data_Historian	0.924
Work_Station	1.000
Corporate_Network	0.867
Sup_PLC	0.226
Sup_PLC_Slave	0.224
Sup_Control_Server	0.226
Sup_HMI	0.240
Sup_Firewall	0.859
Sup_PLC_Master	0.219
Sup_Data_Server	0.234
Sup_Data_Historian	0.867
Sup_Work_Station	0.859
Sup_Corporate_Network	0.859
Sup_PLC_top	0.867

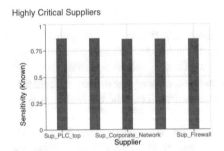
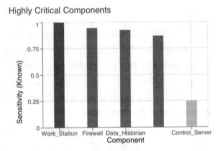

Fig. 5.10 Top five components and suppliers by Birnbaum importance

the supply chain in IoT-enabled infrastructure systems. The tool uses a customized approach to analyze risks in networked systems, such as the IoT, that emanates from the suppliers of individual components. Case studies of an autonomous vehicle and an industrial control system cyber security have been provided to illustrate the risk analysis framework and risk mitigation decisions in Chaps. 2 and 3, respectively.

The supply chain risk management in IoT is a highly convoluted problem to solve. This book has provided an introduction to modeling, computations, and software tools for IoT supply chain risk analysis and mitigation. It has opened up a new dimension in traditional supply chain risk management by understanding vendor involvement and risk propagation in complex and networked systems. Going forward, we envision that advances will be made on several different fronts. The

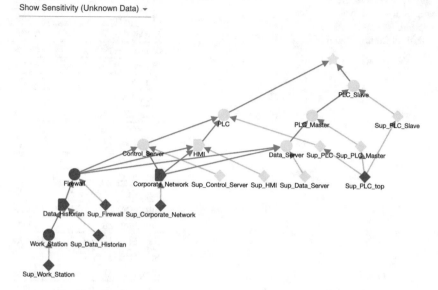

Fig. 5.11 Components and suppliers colored according to Birnbaum structural importances using the scheme shown in Fig. 5.1

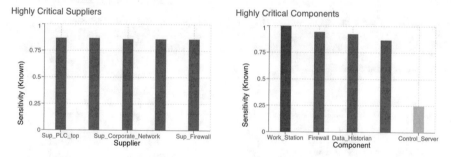

Fig. 5.12 Top five components and suppliers by Birnbaum structural importance in Case Study 2

first direction is the accurate mapping of the threat landscape in terms of the supply chain. Then, a comprehensive risk assessment and impact analysis need to be done. Finally, mitigation strategies are needed to act as a guideline for establishing best practices. Effective solutions to ensure the supply chain security of the IoT require a long trajectory of development. There is a need for active public-private engagement to come up with concrete solutions. Joint policy and technical solutions are needed to counter the risks. Technology should inform policy and policy should regulate technology. Continuous assessment of risks presented by existing suppliers and response strategies is required for an effective defense against the emerging supply chain threats in IoT systems.

Table 5.4 Birnbaum
structural importance
measure for Case Study 2:
Industrial Control System

PLC	0.496
PLC_Slave	0.496
HMI	0.496
Control_Server	0.496
Firewall	1.000
PLC_Master	0.496
Data_Server	0.496
Data_Historian	1.000
Work_Station	1.000
Corporate_Network	1.000
Sup_PLC	0.496
Sup_PLC_Slave	0.496
Sup_Control_Server	0.496
Sup_HMI	0.496
Sup_Firewall	1.000
Sup_PLC_Master	0.496
Sup_Data_Server	0.496
Sup_Data_Historian	1.000
Sup_Work_Station	1.000
Sup_Corporate_Network	1.000
Sup_PLC_top	1.000

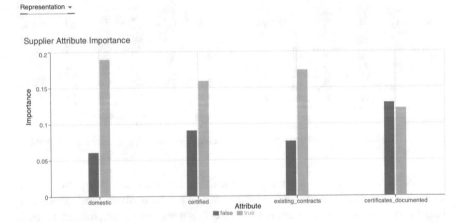

Fig. 5.13 Supplier attributes representation

Since the system-level analysis and decision-making of IoT supply chain security involves high computational complexity, future work can study the use of heuristics to achieve feasibly optimized risk mitigation. Furthermore, it can also consider the interaction of event and supplier risks as well as considering more complex supplier network topologies. A particularly troubling aspect of supply chain security is that

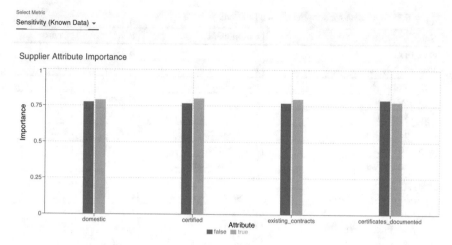

Fig. 5.14 Supplier attributes importances

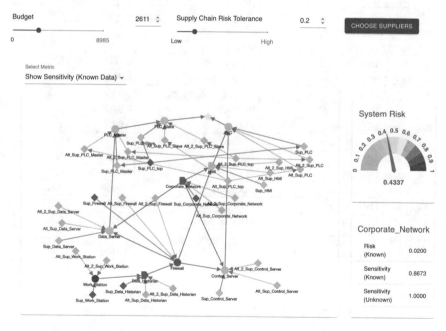

Fig. 5.15 Decision Support panel featuring the Industrial Control System graph

the supply chain may change after a system has been designed and manufactured. In the terms of our analysis in this book, group memberships can be modified after the supplier choices are made. Because many suppliers have ongoing access to systems in the form of maintenance or support access, this dynamism in the supplier network can significantly alter the perceived risk to system security. Hence, more dynamic and real-time risk management techniques can be explored in future extensions of this book.

Table 5.5 Input data for supplier choice problem

Supplier	Component	Risk	Cost
Sup_PLC	PLC	0.08	10
Sup_PLC_Slave	PLC_Slave	0.0700	10
Sup_Control_Server	Control_Server	0.1100	30
Sup_HMI	HMI	0.0900	20
Sup_Firewall	Firewall	0.1000	300
Sup_PLC_Master	PLC_Master	0.0900	100
Sup_Data_Server	Data_Server	0.1100	500
Sup_Data_Historian	Data_Historian	0.0800	250
Sup_Work_Station	Work_Station	0.1500	350
Sup_Corporate_Network	Corporate_Network	0.0200	1000
Alt_Sup_PLC	PLC	0.1600	5
Alt_Sup_PLC_Slave	PLC_Slave	0.1400	5
Alt_Sup_Control_Server	Control_Server	0.2200	15
Alt_Sup_HMI	HMI	0.1800	10
Alt_Sup_Firewall	Firewall	0.2000	150
Alt_Sup_PLC_Master	PLC_Master	0.1800	50
Alt_Sup_Data_Server	Data_Server	0.2200	250
Alt_Sup_Data_Historian	Data_Historian	0.1600	125
Alt_Sup_Work_Station	Work_Station	0.3000	175
Alt_Sup_Corporate_Network	Corporate_Network	0.0400	500
Alt_2_Sup_PLC	PLC	0.0400	20
Alt_2_Sup_PLC_Slave	PLC_Slave	0.0350	20
Alt_2_Sup_Control_Server	Control_Server	0.0700	60
Alt_2_Sup_HMI	HMI	0.0450	30
Alt_2_Sup_Firewall	Firewall	0.0550	600
Alt_2_Sup_PLC_Master	PLC_Master	0.0450	200
Alt_2_Sup_Data_Server	Data_Server	0.0550	1000
Alt_2_Sup_Data_Historian	Data_Historian	0.0400	500
Alt_2_Sup_Work_Station	Work_Station	0.0750	700
Alt_2_Sup_Corporate_Network	Corporate_Network	0.0100	2000

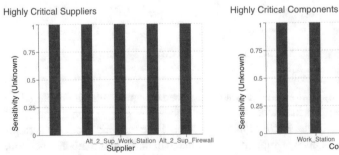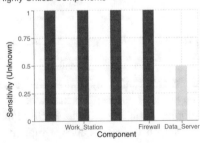

Fig. 5.16 Top five components and suppliers after a low budget optimization

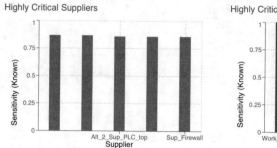
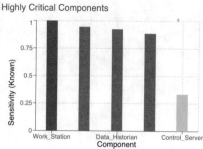

Fig. 5.17 Top five components and suppliers after a high budget optimization

References

1. A. Rauzy, Toward an efficient implementation of the MOCUS algorithm. IEEE Trans. Reliab. **52**(2), 175–180 (2003)
2. R.E. Bryant, Graph-based algorithms for boolean function manipulation. IEEE Trans. Comput. **100**(8), 677–691 (1986)
3. A. Rauzy, New algorithms for fault trees analysis. Reliab. Eng. Syst. Saf. **40**(3), 203–211 (1993)
4. A. Rauzy, Binary decision diagrams for reliability studies, in *Handbook of Performability Engineering* (Springer, 2008), pp. 381–396
5. X. Zang, H. Sun, K.S. Trivedi, A BDD-based algorithm for reliability graph analysis, Department of Electrical Engineering, Duke University, Tech. Rep. (2000)
6. W.S. Jung, S.H. Han, J. Ha, A fast BDD algorithm for large coherent fault trees analysis. Reliab. Eng. Syst. Saf. **83**(3), 369–374 (2004)
7. M.L. Bynum, G.A. Hackebeil, W.E. Hart, C.D. Laird, B.L. Nicholson, J.D. Siirola, J.P. Watson, D.L. Woodruff, *Pyomo–Optimization Modeling in Python*, 3rd edn., vol. 67 (Springer Science & Business Media, 2021)
8. W.E. Hart, J.P. Watson, D.L. Woodruff, Pyomo: modeling and solving mathematical programs in python. Math. Program. Comput. **3**(3), 219–260 (2011)
9. P. Bhavsar, K. Dey, M. Chowdhury, P. Das, Risk analysis of autonomous vehicles in mixed traffic streams, Region 2 University Transportation Research Center (UTRC), New York, NY, Tech. Rep. (2017). [Online]. Available: https://ntlrepository.blob.core.windows.net/lib/62000/62500/62527/Final-Report-Risk-Analysis-of-Autonomous-Vehicles.pdf

Index